Synthesis Lectures on Mathematics & Statistics

This series includes titles in applied mathematics and statistics for cross-disciplinary STEM professionals, educators, researchers, and students. The series focuses on new and traditional techniques to develop mathematical knowledge and skills, an understanding of core mathematical reasoning, and the ability to utilize data in specific applications.

Iordanka Panayotova • Maila Hallare •
Viktoria Savatorova

Mathematical Modeling in Life Sciences

A Practical Project-Based Approach

Iordanka Panayotova
Department of Mathematics
Christopher Newport University
Newport News, VA, USA

Maila Hallare
Department of Mathematics
United States Air Force Academy
Air Force Academy, CO, USA

Viktoria Savatorova
Department of Mathematical Sciences
Central Connecticut State University
New Britain, CT, USA

ISSN 1938-1743 ISSN 1938-1751 (electronic)
Synthesis Lectures on Mathematics & Statistics
ISBN 978-3-032-17485-7 ISBN 978-3-032-17486-4 (eBook)
https://doi.org/10.1007/978-3-032-17486-4

This Springer imprint is published by the registered company Springer Nature Switzerland AG
The registered company address is: Gewerbestrasse 11, 6330 Cham, Switzerland

We dedicate this book to our families for their constant support and unconditional love.

Preface

This book introduces students to mathematical modeling using real-world applications of ordinary differential equations (ODEs) in biology, medicine, and ecology. Built around engaging projects that challenge students to work with both real datasets and theoretical scenarios, it provides instructors with ready-to-use modules that emphasize model construction, analysis, and interpretation, supported by MATLAB code and student exercises. This structure reflects the broader goal of the book: to teach modeling with differential equations, systems of differential equations, and data-driven dynamical systems in the context of life sciences, offering a comprehensive range of modeling techniques. Our project-based approach ensures that students actively engage with the material through contemporary case studies, real data, and hands-on exploration. Students also acquire essential computational skills through coding and gain experience working with real-world data—skills that are increasingly valuable in today's data-driven world.

A key emphasis is placed on interpreting the biological meaning of results derived from mathematical and computational analysis, so students not only master the technical tools but also gain an appreciation of how these tools inform scientific discovery. This approach has proven highly effective in our own classrooms, where students have responded enthusiastically to real-world applications and consistently reported greater motivation and confidence in end-of-semester reflections. This book bridges the gap between theory and practice by equipping students with the tools needed to become confident modelers, researchers, and practitioners in the life sciences.

The volume is intended as a textbook for junior and senior undergraduate students in courses as Mathematical Modeling, Advanced Differential Equations, or Dynamical Systems. Chapter 1 provides a broad overview of the entire mathematical modeling process and explains why modeling is such a powerful tool for understanding complex systems in life sciences. Students are introduced to the idea that models serve as simplified representations of reality—balancing assumptions, structure, and data to generate insight, predictions, and potential decision-making tools.

Chapter 2 deals with modeling diabetes onset and management through linear and nonlinear models. In Sect. 2.2, we use the Laplace transform to solve a linear system modeling self-management of type 2 diabetes. We also introduce curve fitting methods to approximate patient-specific time-series data for blood glucose and hormone concentrations using a limited number of parameters. Section 2.3 applies optimal control theory to determine an insulin infusion strategy that keeps glucose levels near a desired target while minimizing total insulin usage. Section 2.4 explores diabetes onset as a bifurcation phenomenon in a nonlinear model. We analyze equilibrium points, their stability, and conduct parameter sensitivity analysis to identify a critical bifurcation parameter.

Chapter 3 focuses on ecological applications, with particular emphasis on multi-species predator–prey dynamics in the Chesapeake Bay. Section 3.2 begins with logistic growth modeling of a single-species population, while Sect. 3.3 introduces predator–prey models and demonstrates how they can be fitted to observational data. Section 3.4 is devoted to parameter sensitivity analysis (both local and global), providing a framework to evaluate model robustness by examining how small changes in system parameters influence solution behavior over time. Building on these foundations, Sect. 3.5 develops a three-species model to study the biological interactions among Atlantic menhaden (prey), striped bass (predator), and the non-native blue catfish (predator/invader). Section 3.6 applies the three-species model to study key ecological challenges, including the overfishing of menhaden, the invasiveness of blue catfish, and harvesting strategies aimed at controlling blue catfish populations. Finally, in Sect. 3.7, we use numerical simulations to study the long-term population dynamics of the three fishes.

Chapter 4 explores cancer–immune interactions and treatment strategies through a series of ODE-based models. It culminates in a melanoma immunotherapy case study, where using the quasi-steady-state reasoning, we obtain and analyze an analytically tractable reduced model. Section 4.2 introduces students to classical growth laws for solid tumors. In Sect. 4.3, we discuss the motivation for modeling tumor–immune interactions and the biological questions such models can address. In Sect. 4.4, we extend these ideas by applying predator–prey framework to model tumor treatment dynamics, illustrating how concepts from ecology can be effectively adapted to oncology. Section 4.5 presents a melanoma case study that couples logistic tumor growth, CTL killing enhanced by antibodies, and antigen-driven recruitment. Section 4.6 is devoted to the numerical simulation and analysis of the reduced tumor–immune competition model, illustrating how computational experiments complement analytical results and provide quantitative insights into tumor–immune dynamics. Finally, Sect. 4.7 connects mathematics back to biology through the hallmarks of cancer framework.

Each chapter is accompanied by student exercises and supported by MATLAB code available on GitHub. Together, these projects link mathematical theory to meaningful practice, helping students develop technical skills while fostering an appreciation of mathematics as a powerful tool for discovery and decision-making in life sciences.

Ultimately, this book is intended to serve as a foundation for exploration. By introducing the essential principles, methods, and applications of mathematical modeling, it equips students with the tools to pursue their own independent projects and investigations.

Whether applied to medicine, ecology, or other areas of the life sciences, the models and ideas developed here are meant to inspire continued inquiry, creativity, and discovery beyond the classroom.

Newport News, VA, USA
Colorado Springs, CO, USA
New Britain, CT, USA
December 2025

Iordanka Panayotova
Maila Hallare
Viktoria Savatorova

Acknowledgments This book grew out of a SQuaRE at the American Institute of Mathematics (AIM). The authors thank AIM for its generous support and for providing an environment that encouraged collaboration and mathematical creativity. I.N. Panayotova gratefully acknowledges Christopher Newport University for granting her a sabbatical year, which provided the dedicated research time needed for the writing of this book. V. Savatorova gratefully acknowledges the reassigned time provided by Central Connecticut State University, which supported the development of this book.

Competing Interests The authors have no competing interests to declare that are relevant to the content of this manuscript.

Contents

Mathematical Modeling 1

1.1 Why We Need Mathematical Modeling?

Mathematical modeling is a powerful tool that allows researchers to translate real-world phenomena into a quantitative framework. By constructing and analyzing models, we can test different assumptions, explore system dynamics, and generate new understanding about complex biological, physical, ecological, or social systems [1, 2]. At its core, modeling is a simplification of real-world phenomena: it involves identifying essential mechanisms by making explicit assumptions and expressing them through mathematical relationships that can be systematically explored and refined. Depending on context, these relationships may be formulated in deterministic or stochastic terms, with each choice carrying interpretive and methodological implications [3, 4].

One of the central strengths of mathematical modeling lies in its ability to test hypotheses through controlled variation of parameters. By adjusting model parameters—such as growth rates, transmission coefficients, or treatment efficacies—researchers can examine how sensitive the outcomes are to these assumptions. This systematic exploration not only helps identify which factors have the greatest influence on system behavior but also reveals interactions that might be obscured in purely observational studies. Good examples of systematic parameter variation and sensitivity analysis can be found in [5] and in the pedagogical treatment by [6].

Models also play a crucial role in hypothesis generation and experimental design. They can suggest relationships or mechanisms that are not immediately evident from empirical data and thus help guide laboratory or clinical experiments [7]. For example, in biomedical research, models of oncolytic virotherapy can identify which viral or tumor properties most strongly determine treatment success and thereby prioritize experimental variants [8]; likewise, tumor–immune dynamical models have been used to propose cytokine-dosing strategies and to anticipate microenvironmental constraints before testing in vivo [9]. In

I. Panayotova et al., *Mathematical Modeling in Life Sciences*, Synthesis Lectures on Mathematics & Statistics, https://doi.org/10.1007/978-3-032-17486-4_1

this way, mathematical modeling provides a bridge between theory and experiment—offering informed, testable hypotheses that can accelerate discovery [7, 10].

Beyond specific applications, mathematical models exemplify the broader power of mathematics to explain the mechanisms underlying biological, physical, and economic systems [11, 12]. By linking equations to data, models reveal structure and coherence in complex phenomena and provide a language through which interdisciplinary understanding can develop [13]. Modeling transforms intuition into quantifiable reasoning, allowing scientists to predict, explain, and visualize processes across scales—from molecular interactions to population-level dynamics [11, 14].

Importantly, models can address questions that are difficult or even impossible to study directly through experiments. In medicine, ethical, logistical, or temporal constraints often limit the ability to perform extensive empirical trials. Mathematical models serve as virtual laboratories, enabling researchers to perform *in silico* experiments, test alternative scenarios, and explore long-term outcomes that would otherwise be inaccessible [15–17].

A similar advantage is found in *ecological and population dynamics modeling*, where the relevant timescales—spanning years or even decades—make direct experimentation impractical. Models of interacting species, nutrient cycles, or climate-driven systems allow scientists to examine how populations evolve under environmental variation, competition, or harvesting pressure [18–21]. Such models not only reveal emergent behaviors—oscillations, extinction thresholds, and coexistence equilibria—but also provide a basis for developing *management and conservation strategies*, such as setting sustainable harvest limits, predicting the effects of invasive species, or assessing ecosystem resilience under climate change [22–24]. In both biomedical and ecological contexts, mathematical models extend the reach of experimentation, offering a framework for hypothesis testing, long-term forecasting, and evidence-based decision-making.

In the context of personalized medicine, mathematical modeling of treatment responses represents an especially promising approach [7, 25]. By predicting likely outcomes of different therapeutic protocols, models can help narrow the range of possibilities that need to be tested experimentally—using virtual patients or *in silico* trial frameworks to prioritize regimens and dosing strategies [16, 26]. This modeling-guided approach reduces cost, saves time, and increases the likelihood of clinical success [16].

Ultimately, the modeling process provides a rigorous, iterative pathway—from conceptualization and abstraction, through computation and simulation, to interpretation and application—advancing both scientific understanding and practical decision-making [15]. Because mathematical modeling plays a central role in understanding, predicting, and managing complex biological and physical systems, it is essential to teach students how to construct, analyze, and validate models themselves. Doing so equips them not only with quantitative reasoning skills but also with a structured way of thinking about dynamic systems—bridging theory, data, and real-world application.

1.2 Modeling Process

The process of mathematical modeling typically unfolds through a series of interconnected stages that form an iterative cycle. A visual illustration of these stages is represented in Fig. 1.1. It begins with analysis of available data and problem formulation, where the real-world system is described qualitatively, relevant variables are identified, and simplifying assumptions are made to capture the system's essential features. The next step is model construction, in which these conceptual relationships are translated into mathematical expressions. Throughout this book we adopt an ODE-centered approach, representing dynamics by systems of ordinary differential equations (ODEs) posed as initial-value problems,

$$\frac{dX}{dt} = F(X, \theta, t), \qquad X(t_0) = X_0, \tag{1.1}$$

where t is the independent variable, X is the vector of dependent variables and θ is the vector of parameter values, and then developing tools within this framework.

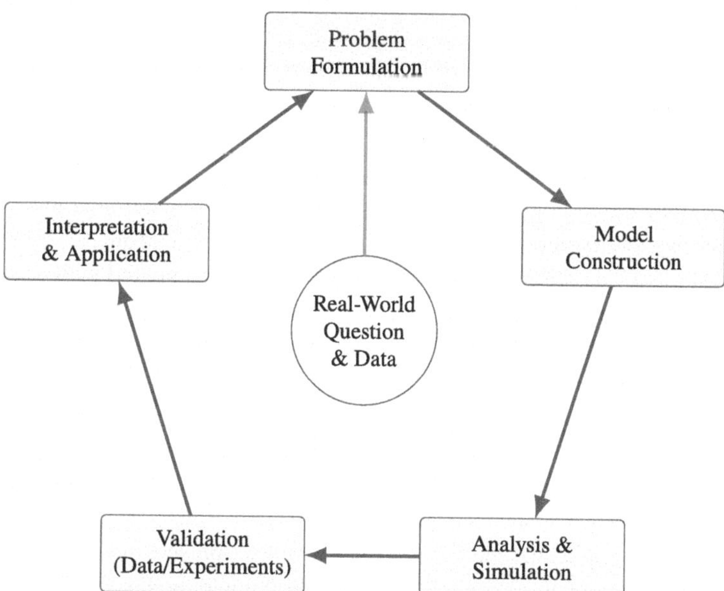

Fig. 1.1 A schematic illustration of the iterative process of mathematical modeling. Beginning from a real-world question, the cycle proceeds through problem formulation, model construction, analysis and simulation, validation against data, and interpretation and application, highlighting continuous model refinement

The ODE-models are typically grounded in underlying biological principles—such as physiological regulation in diabetes, tumor–immune interactions in cancer, or predator–prey dynamics in ecological systems—although in many cases they may also be developed from data-driven formulations. It is important to recognize that every mathematical model reflects the assumptions on which it is built. Modifying these assumptions—whether by relaxing simplifying constraints or incorporating additional mechanisms—can lead to models with different parameter values, altered functional forms, or even new state variables. Such changes may affect not only the quantitative outcomes but also the qualitative behavior of the system, potentially introducing new steady-states, bifurcations, or dynamical regimes. These ideas are explored in subsequent chapters through models of glucose–insulin regulation, cancer–immune interactions, and multi-species dynamics in the Chesapeake Bay, demonstrating how different modeling assumptions lead to different system behaviors and interpretations.

Once a model is constructed, it is subjected to analysis and numerical simulations. Analytical techniques, numerical methods, and computational tools are usually employed to explore the model's behavior, derive steady states, assess stability, and generate quantitative predictions. Some of the analytical techniques we employ include deriving equilibria and performing linearization to assess local stability, applying Lyapunov's direct method to establish global stability, and conducting bifurcation analysis to characterize qualitative changes in system dynamics under parameter variation. In practical applications, we rely on parameter estimation techniques, such as the least squares method, to infer parameter values that best align the model with observed data, enabling the extraction of biologically meaningful information from empirical observations. This step also includes sensitivity analysis to determine how variation in parameters influences outcomes and to identify which assumptions have the greatest impact on model results [27, 28]. We discuss both local and global sensitivity analysis, including PRCC (partial rank correlation coefficients [29]) and variance-based indices (e.g., Sobol' S_i, S_{T_i} [30]), to determine how parameter variability propagates to outputs and to identify influential mechanisms.

The next stage is validation, where model predictions are compared with experimental, observational, or empirical data. Any discrepancies between simulated and observed outcomes are not model "failures" but valuable opportunities for model refinement which could be done by revising the assumptions, parameter recalibration, or including additional mechanisms (and, when warranted, new parameters). This iterative process of testing and improvement enhances both the model's predictive accuracy and its biological interpretability, ensuring that the mathematical formulation remains grounded in empirical reality.

Although more complicated models can, in principle, represent biological reality more faithfully, greater complexity is not always desirable. Simple models often yield valuable insight: they are analytically tractable, easier to interpret, and less prone to overfitting—especially when data are limited. Parsimonious (*in modeling it means simplicity without unnecessary complexity*) formulations also improve parameter identifiability and make uncertainty analysis more transparent. This perspective aligns with **Occam's (Ockham's)**

principle: *when two models explain the data equally well, one should choose the model with fewer assumptions and parameters* [31, 32]. In practice, we operationalize this by building model hierarchies (from simple to complex), advancing only when added complexity significantly improves fit or predictive power of the model. When comparing two similar models, model parsimony can be formalized using information criteria such as the **Akaike Information Criterion (AIC)** and the **Bayesian Information Criterion (BIC)**, which quantify the trade-off between model fit and complexity.

The AIC criteria was introduced by Hirotugu Akaike [33] and is defined as

$$AIC = 2k - 2\ln(L),$$

where k is the number of parameters and L is the maximized likelihood. The BIC or **Schwarz criterion**, derived by Gideon Schwarz in 1978 [34] is defined as

$$BIC = k\ln(n) - 2\ln(L),$$

where n is the number of observations. For both criteria, lower values indicate a better balance between fit and parsimony. When two models explain the data comparably well, selecting the one with lower AIC or BIC complies with the Occam's principle and helps guard against overfitting while preserving interpretability.

Finally, the interpretation and application stage connects the mathematics back to the real-world context. The model's outcomes are translated into biological, physical, medical or policy insights that can inform decision-making, suggest further experiments, or guide management strategies. In this way, modeling is not a linear task but a continuous dialogue between mathematics and reality—one that deepens understanding, drives innovation, and supports evidence-based reasoning across disciplines.

Hence, we should teach students not only how to construct and analyze models but also how to interpret their results, communicate their findings effectively, and connect their conclusions to the real-world phenomena that motivated the modeling in the first place. Cultivating these skills ensures that mathematical modeling serves not only as a tool for discovery but also as a framework for meaningful scientific communication, interdisciplinary collaboration, and evidence-based decision-making. In later chapters, we illustrate these principles through projects on glucose–insulin regulation, cancer–immune dynamics, and multi-species ecological systems, where students learn to interpret model outcomes in context and communicate conclusions grounded in both data and theory.

1.3 Parameter Estimation and Model Calibration

Mathematical models become even more informative when their parameters are estimated from real-world observations. Parameter estimation—also referred to as model calibration—is the process of determining parameter values that enable a model to reproduce observed behaviors of the system it represents. In this stage, theory and data intersect: while the equations describe the hypothesized mechanisms, data provide the quantitative evidence needed to specify how fast or how strongly these mechanisms act.

For example, in a 2×2 linear ODE model of blood-glucose regulation used in Chap. 2 to introduce diabetes modeling, parameters such as glucose effectiveness, insulin sensitivity, and basal production/clearance rates are estimated from *patient-specific time-series data.* The goal is to find the parameter set that minimizes the discrepancy between model predictions and observed glucose (and, when available, insulin) measurements. One of the most commonly used approaches is the *least squares* method, which seeks to minimize the sum of squared residuals between data and model output:

$$\min_{\boldsymbol{\theta}} \sum_{i=1}^{n} [X_{\text{data}}(t_i) - X_{\text{model}}(t_i; \boldsymbol{\theta})]^2 ,$$

where θ denotes the vector of parameters to be estimated.

An important practical question in data fitting is *how to choose initial parameter values for the optimization.* Nonlinear least squares methods rely on iterative algorithms that start from an initial guess and refine the parameters to minimize residuals. Poor initial guesses can lead to slow convergence or convergence to local, rather than global, minima. In practice, initial values can be informed by prior knowledge, experimental measurements, literature values, or simple approximations derived from the data itself. For example, linearizing part of the model or considering limiting cases can provide reasonable starting points. Exploring multiple initial guesses is often helpful to ensure robust parameter estimation.

A critical consideration during calibration is parameter identifiability. Even if a model fits the data well, not all parameters may be uniquely determined. Structural identifiability concerns whether unique parameter values exist in principle given perfect, noise-free, continuous data, whereas practical identifiability addresses whether the available data contain sufficient information to reliably estimate those parameters [35, 36]. Models with too many parameters relative to the data risk overfitting (fitting noise rather than true biological patterns) and may lose predictive power; structural identifiability is a necessary but not sufficient condition for reliable estimation [37]. To mitigate this risk, one can compare competing models using information criteria such as the Akaike Information Criterion or Bayesian Information Criterion, which penalize model complexity and codify the trade-off between fit and parsimony [38].

In this book, parameter estimation plays a central role in linking differential equation models to real-world data. Through case studies involving population dynamics and biomedical systems, we illustrate how simple fitting procedures—such as least squares optimization—can extract biologically meaningful parameters, quantify uncertainty, and guide further model refinement. Calibration transforms mathematical models from abstract representations into quantitative predictive tools.

1.4 Model Analysis and System Behavior

Once a model has been formulated and its parameters estimated, the next step is to analyze its behavior. Model analysis provides insight into the qualitative and quantitative properties of the system—how it evolves over time, responds to perturbations, and approaches steady states if they exist. Through analysis, we can move beyond reproducing data to understanding why certain outcomes occur and how they depend on key parameters.

In the context of ordinary differential equation (ODE) models, analysis typically begins with identifying the equilibrium points (or steady states) of the system—values of the variables where the rates of change vanish. These equilibria may represent biologically meaningful conditions such as disease-free, coexistence, or extinction states. Determining their existence and feasibility provides a first step toward understanding the long-term dynamics of the system.

Because the systems under study are usually complex and involve interacting components, we study their behavior near an equilibrium by linearizing the system, replacing the full nonlinear model with a simpler linear approximation. The resulting Jacobian matrix captures how small deviations from equilibrium evolve. The eigenvalues of this matrix determine the local stability of the equilibrium: if all eigenvalues have negative real parts, perturbations decay, and the system returns to equilibrium. Conversely, if any eigenvalue has a positive real part, the equilibrium is unstable and the system diverges toward another state.

While local analysis provides valuable information about the behavior near equilibria, many biological systems exhibit nonlinear interactions that can lead to complex global behavior. To capture these dynamics, we extend the analysis using Lyapunov stability theory, which allows us to assess global stability without solving the equations explicitly. By constructing a suitable Lyapunov function, one can demonstrate that solutions converge toward a stable steady state for a wide range of initial conditions. Such analyses are particularly relevant in biomedical models, where stability of a tumor-free or infection-free equilibrium can correspond to successful therapy or immune control.

Beyond stability, models can exhibit qualitative transitions in behavior as parameters vary—a phenomenon known as bifurcation. Bifurcation analysis identifies critical thresholds beyond which system behavior changes, such as the transition from tumor dormancy to growth, or from coexistence to extinction in ecological systems. These threshold phenomena are central to understanding the sensitivity and resilience of complex biological systems.

In addition to analyzing natural system behavior, we may also seek to guide or optimize a system's trajectories. *Optimal control methods* provide a mathematical framework for designing strategies that achieve desired objectives—such as regulating blood glucose, or maintaining ecological balance—while respecting the system's intrinsic dynamics and constraints.

Together, *equilibrium*, *stability*, and *bifurcation* analyses reveal how system behavior emerges from model structure and parameter values. These mathematical tools, introduced here conceptually, will be developed in detail in later chapters and applied to concrete biological and biomedical examples. They provide the foundation for interpreting model dynamics, uncovering feedback mechanisms, linking mathematical results with observable outcomes, and ultimately designing effective interventions through optimal control.

1.5 Summary

Mathematical modeling provides a *systematic framework* for translating real-world phenomena into equations that can be analyzed, simulated, and interpreted. Throughout this chapter, we have emphasized that modeling is *an iterative process*—beginning with conceptualization and abstraction, followed by model construction, analysis, calibration, and validation. Each stage strengthens the connection between mathematics and the system it represents, allowing us to explain, predict, and refine our understanding of complex natural and biomedical processes.

A central message is the *balance* between realism and simplicity. While complex models can represent biological detail more faithfully, simpler models often yield clearer insights and permit analytical treatment. By focusing on key mechanisms and estimating parameters from data, even minimal models can capture essential features of a system's behavior and provide predictive and explanatory power.

Two core components—*parameter estimation* and *model analysis*—serve as the methodological pillars of this book. Parameter estimation connects models to data, grounding them in measurable biological or ecological quantities. Model analysis, in turn, reveals how these models behave dynamically, identifying equilibria, stability conditions, and qualitative transitions. Together, these steps transform equations into interpretable tools for understanding real phenomena.

The modeling philosophy introduced in this chapter will be applied throughout the book to a variety of systems that, despite their differences, share a common mathematical structure. In modeling diabetes, we examine physiological regulation and therapeutic intervention through feedback control of glucose–insulin dynamics. In cancer–immune interaction models, we explore how nonlinear feedback between tumor and immune populations shapes treatment outcomes and stability of remission. Finally, in multi-species ecological models of the Chesapeake Bay, we study predator–prey and competitive interactions that govern ecosystem balance and resilience under environmental change.

Across these case studies (diabetes, cancer, and population dynamics) we see an unifying principle: the same mathematical language of ordinary differential equations can describe processes across scales, from cells to organisms to ecosystems. More importantly, these examples show how modeling connects data, mathematics, and interpretation, enabling us to understand complex systems, generate hypotheses, and support evidence-based decisions. This chapter has introduced the foundational ideas and workflow of mathematical modeling, setting the stage for the following chapters, where ODEs will be applied in depth to ecological and biomedical problems.

References

1. Ingalls, B.P., Mathematical Modeling in Systems Biology: An Introduction. MIT Press, Cambridge, MA (2013)
2. Torres, N.V., Santos, G., The (Mathematical) Modeling Process in Biosciences. *Front. Genet.*, **6**, 354 (2015)
3. Korsbo, N., Jönsson, H., It's about time: Analysing simplifying assumptions for modelling multi-step pathways in systems biology. *PLoS Comput. Biol.*, **16**(6), e1007982 (2020)
4. Wilkinson, D.J., Stochastic Modelling for Systems Biology. Chapman & Hall/CRC, Boca Raton, FL (2006)
5. Drechsler, M., Sensitivity analysis of complex models. *Biological Conservation* **86**(3), 401–412 (1998)
6. Savatorova, V., Exploring parameter sensitivity analysis in mathematical modeling with ordinary differential equations. *CODEE Journal* **16**, Article 4 (2023)
7. Butner, J.D., Dogra, P., Chung, C., Pasqualini, R., Arap, W., Lowengrub, J., Cristini, V., Wang, Z., Mathematical modeling of cancer immunotherapy for personalized clinical translation. *Nat. Comput. Sci.*, **2**(12), 785–796 (2022)
8. Guo, E., Dobrovolny, H.M., Mathematical Modeling of Oncolytic Virus Therapy Reveals Role of the Immune Response. *Viruses*, **15**(9), 1812 (2023)
9. Harkos, C., Stylianopoulos, T., Jain, R.K. Mathematical modeling of intratumoral immunotherapy yields strategies to improve the treatment outcomes. *PLoS Comput. Biol.*, **19**(12), e1011740 (2023)
10. Panayotova, I.N., Hallare, M., Oncolytic Virus Versus Cancer: Modeling and Simulation of Virotherapy with Differential Equations. In: *New Trends in the Applications of Differential Equations in Sciences*. Springer (2023). doi:10.1007/978-3-031-21484-4_23
11. Murray, J.D., Mathematical Biology I: An Introduction. 3rd edn. Springer, New York (2002)
12. Strogatz, S.H., Nonlinear Dynamics and Chaos: With Applications to Physics, Biology, Chemistry, and Engineering. 2nd edn. Westview Press, Boulder (2015)
13. Brunton, S.L., Kutz, J.N., Data-Driven Science and Engineering: Machine Learning, Dynamical Systems, and Control. Cambridge University Press, Cambridge (2019)
14. Otto, S.P., Day, T., A Biologist's Guide to Mathematical Modeling in Ecology and Evolution. Princeton University Press, Princeton (2007)
15. Winslow, R.L., Trayanova, N., Geman, D., Miller, M.I. Computational Medicine, Translating Models to Clinical Care. *Sci. Transl. Med.* **4**(158), 158rv11 (2012)
16. Pappalardo, F., Russo, G., Musuamba Tshinanu, F., Viceconti, M., In silico clinical trials: concepts and early adoptions. *Brief. Bioinform.* **20**(5), 1699–1708 (2019)
17. Metzcar, J., Wang, Y., Heiland, R., Macklin, P.: A Review of Cell-Based Computational Modeling in Cancer Biology. *JCO Clin. Cancer Inform.* **3**, 1–13 (2019)
18. Panayotova, I. N., Hallare, M., Modeling the Ecological Dynamics of a Three-Species Fish Population in the Chesapeake Bay. CODEE Journal **14**, Article 2 (2021). https://doi.org/10.5642/codee.202114.01.02
19. Panayotova, I.N., Hallare, M., Modeling ecological issues in fish dynamics: Competing predators, invasiveness, and over-harvesting. Ecological Modelling **466**, 109885 (2022)
20. May, R.M.: Simple mathematical models with very complicated dynamics. Nature **261**, 459–467 (1976)
21. Murray, J.D., Mathematical Biology: An Introduction. 3rd edn. Springer, New York (2002)
22. Clark, C.W., Mathematical Bioeconomics: The Mathematics of Conservation. 3rd edn. Wiley, Hoboken, NJ (2010)

23. Hallare, M., Panayotova, I.N.: Maintaining Ecosystem and Economic Structure in a Three-Species Dynamical System in Chesapeake Bay. CODEE J. **15**, Article 3 (2022). https://doi.org/10.5642/codee.202215.01.03
24. Panayotova, I.N., Herrmann, J., Kolling, N. (2023) Bioeconomic analysis of harvesting within a predator-prey system: A case study in the Chesapeake Bay fisheries. *Ecological Modeling*, 480, 110330.
25. Kondic, J., Jarrett, J., Keener, J.P.: The Use of Mathematical Modeling in Preclinical Decision Making. Comput. Syst. Oncol. **2**(2), e12107 (2022)
26. Craig, M., Thomas, R., et al., A practical guide for the generation of model-based virtual patients and virtual trials. Front. Syst. Biol. **3**, 1174647 (2023)
27. Saltelli, A., Chan, K., Scott, EM. *Sensitivity analysis*, Wiley series in probability and statistics. Wiley, Chichester; New York. (2000)
28. Saltelli, A., Ratto, M., Andres, T., Campolongo, F., Cariboni, J., Gatelli, D., Saisana, J., Tarantola, S. *Global Sensitivity analysis: The primer*, Wiley, LTD., Chichester. (2008)
29. Kendall, MG. (1942) Partial rank correlation. *Biometrica*, 32(3/4): 277–283.
30. Sobol', IM. (1993) Sensitivity estimates for nonlinear mathematical models. *Math. Model. Comput.Exp.*, 1(4):407–414.
31. Burnham, K.P., Anderson, D.R.: Model Selection and Multimodel Inference: A Practical Information-Theoretic Approach. 2nd edn. Springer, New York (2002)
32. Sober, E.: Ockham's Razors: A User's Manual. Cambridge University Press, Cambridge (2015)
33. Akaike, H. Information theory as an extension of the maximum likelihood principle. In: Petrov, BN and Csaki, F. In *Second International Symposium on Information Theory*. Akademiai Kiado, Budapest, pp. 276–281. (1973)
34. Schwarz, G. Estimating the Dimension of a Model. *Ann. Statist.* **6** (2) 461–464, March, (1978). https://doi.org/10.1214/aos/1176344136.
35. Wieland, F.-G., Hauber, A.L., Rosenblatt, M., Tönsing, C., Timmer, J.: On structural and practical identifiability. Curr. Opin. Syst. Biol. **25**, 60–69 (2021)
36. Structural and practical identifiability analysis in bioengineering. J. Biol. Eng. **18**, 26 (2024)
37. Mentre, F., Mallet, A., Baccar, D.: Parameter identifiability of fundamental pharmacodynamic models. Front. Physiol. **7**, 590 (2016)
38. Dziak, J.J., Coffman, D.L., Lanza, S.T., Li, R., Jermiin, L.S.: Sensitivity and specificity of information criteria. Brief. Bioinform. **21**(2), 553–565 (2020)

Modeling Diabetes 2

2.1 Introduction

Diabetes is a major. and growing public health concern in the United States. An estimated 38.4 million Americans (about 11.6% of the population) have diabetes, accounting for roughly two-thirds of all cases across developed countries [1]. Among these, approximately 1 in 5 remain undiagnosed. Additionally, over 96 million U.S. adults are estimated to have prediabetes, yet more than 80% of them are unaware of their condition [2].

Diabetes is associated with a range of serious health complications. For instance, among adults diagnosed with diabetes, 69% also have high blood pressure, 44% have high cholesterol, and 39% have chronic kidney disease [2]. There is also increasing evidence linking diabetes to elevated cancer risk and worse cancer-related outcomes [3].

In 2022, the Center for Disease Control and Prevention released new findings highlighting socioeconomic disparities in diabetes prevalence, showing that individuals below the federal poverty level are disproportionately affected [2]. These trends emphasize the need for broader educational and scientific efforts to understand and manage the disease.

There are three main types of diabetes: type 1, type 2, and gestational diabetes. Gestational diabetes can develop during pregnancy in otherwise healthy women and usually resolves after delivery. Type 1 diabetes is an autoimmune disease that generally manifests in children, adolescents, and young adults, but can occur at any age. It is called "insulin-dependent" diabetes, as if a patient has type 1 diabetes, their pancreas no longer produces the insulin needed to balance blood sugar levels and no lifestyle changes can reverse this condition. As a result, patients become dependent on insulin for the rest of their lives. Type 2 diabetes is the most common type of diabetes (95% of all cases), and this is the reason why our next discussion will focus on it. Type 2 diabetes typically develops in individuals over the age of 45 years, although it can also occur in younger adults. Often referred to as "insulin resistant" diabetes, this condition arises when the pancreas does not

I. Panayotova et al., *Mathematical Modeling in Life Sciences*, Synthesis Lectures on Mathematics & Statistics, https://doi.org/10.1007/978-3-032-17486-4_2

produce enough insulin or the body becomes resistant to its effects. Although a healthy lifestyle such as a balanced diet, regular exercise, and weight management can prevent or help to manage type 2 diabetes, administering medications may still be necessary [2].

This chapter explores how mathematical modeling can be used to better understand the development and management of type 1 and type 2 diabetes. By modeling glucose-insulin dynamics and treatment strategies using ordinary differential equations (ODE), we provide tools to analyze physiological processes through a mathematical lens.

Designed for use in undergraduate courses such as Differential Equations, Dynamical Systems, or Mathematical Modeling, this chapter offers scaffolded projects that guide students through building, analyzing, and interpreting mathematical models. The projects emphasize real-world data, biological interpretation, and model-based decision making.

Section 2.2 introduces a linear ODE model of the glucose regulatory system. Parameters are estimated using patient data, treatment and food intake are modeled as discontinuous inputs, and the system is solved using the Laplace transform. Section 2.3 applies optimal control theory to determine an insulin infusion strategy that minimizes insulin use while keeping glucose levels near a target range. Section 2.4 introduces a nonlinear model to study the onset of type 2 diabetes, focusing on equilibrium analysis, stability, and parameter sensitivity. Students explore how changes in key physiological parameters can lead to a bifurcation, triggering a transition from healthy to diabetic dynamics.

Each project is designed to strengthen students' ability to formulate and analyze mathematical models, as well as to obtain and interpret results. Emphasis is placed on the role of model assumptions and the choice of solution methods, with particular attention to the sensitivity of the system to parameter values. MATLAB code is provided to support both analytical and numerical exploration.

By engaging with these models, students gain experience applying differential equations to an important biomedical problem, deepening their understanding of mathematics and the dynamics of underlying physiological processes.

2.2 Linear Models for Diabetes Self-Management

Before we start developing mathematical models, it is useful to examine a simplified schematic of the blood glucose regulatory system, shown in Fig. 2.1. This diagram illustrates how glucose levels are regulated in humans through the interaction of food intake, tissue metabolism, and liver function. As food enters into the gastrointestinal track, it is reduced to glucose and enters the bloodstream. In addition to food intake, the blood glucose concentration can be altered by tissue metabolism and storage in (or release from) the liver. The liver stores glucose in the form of glycogen and releases it back to blood depending upon the body's need. The endocrine gland in the pancreas secrets the hormones insulin and glucagon which regulate the blood glucose level. Glucagon prevents the blood glucose level from dropping too low and signals the liver to break down glycogen and release glucose when a patient does not eat for a long time, especially overnight. Insulin

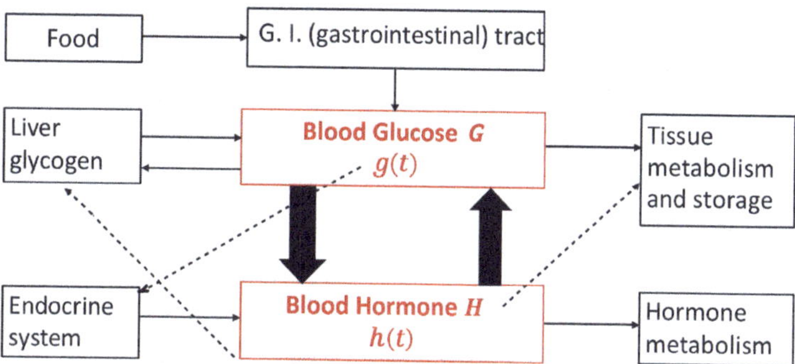

Fig. 2.1 Simplified model of the blood glucose regulatory system in a human

is released in response to rising glucose levels in the bloodstream; an increase in insulin results in increased metabolism of excess hormone. Diabetes manifests itself as Type 1 when insulin is not produced in the pancreas. Type 2 diabetes is a situation when the pancreas can't keep up sufficient insulin production or cells stop responding to signals to produce insulin. Both cases result in an excess of glucose in the bloodstream [4,5].

2.2.1 Model Derivation and Justification

Our goal is to use mathematical modeling to better understand the dynamics of blood glucose regulation, distinguish between individuals with and without diabetes, and test various treatment scenarios for effective diabetes management. For instructional purposes, we simplify the blood glucose regulatory system, focusing on two key quantities: the blood glucose concentration, $G(t)$, and the net hormonal (insulin) concentration, $H(t)$, both varying as functions of time t. The following assumptions are made and are schematically illustrated in Fig. 2.2:

- The liver releases glucose into the bloodstream at a basal (constant background) rate I_b independent of food intake or insulin levels.
- The rate of glucose removal independent of insulin from the digestive system is proportional to the glucose itself with a constant of proportionality a.
- The rate of glucose removal dependent of insulin is proportional to the insulin with a constant of proportionality b.
- The external rate at which the blood glucose concentration increases as a result of the food intake is a function $I = I(t)$ on time.
- The rate at which the pancreas releases insulin into the bloodstream, regardless of blood glucose levels is a constant H_b.
- The average rate of insulin removal from the body independent of glucose is directly proportional to the insulin present with a constant of proportionality c.

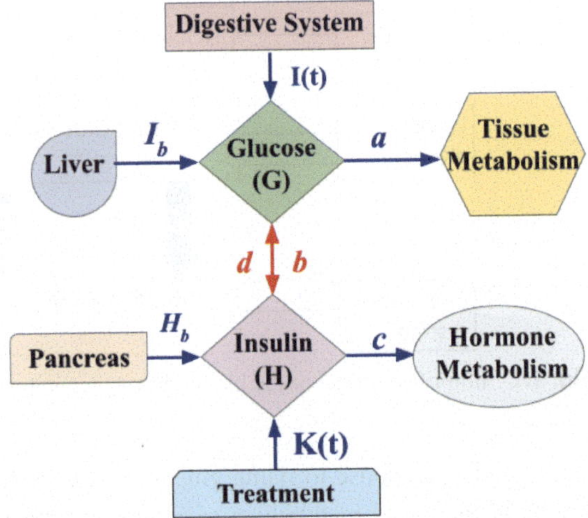

Fig. 2.2 Schematic illustration of the mathematical model of the blood glucose-insulin dynamics

- The rate of insulin release stimulated by glucose is proportional to the current glucose concentration, with a constant of proportionality d. For type 1 diabetes, $d = 0$, reflecting the absence of insulin secretion. For type 2 diabetes, d is positive but smaller than in healthy individuals, representing reduced insulin responsiveness.
- The effect of other hormones such as cortisol, epinephrine, growth hormone, etc. is neglected.

Under these assumptions, the mathematical model of glucose-insulin dynamics is given by the following system of two linear differential equations

$$\begin{cases} \dfrac{dG}{dt} = I_b - aG - bH + I(t) \\ \dfrac{dH}{dt} = H_b - cH + dG. \end{cases} \tag{2.1}$$

This model contains six parameters. To reduce the number of parameters, we consider the system at the fasting state, where no food is being consumed, and the glucose and insulin concentrations remain constant at their fasting levels, denoted by G_f and H_f, respectively. At this steady state, the rates of change of glucose and insulin are both zero, so $\dfrac{dG}{dt} = 0$ and $\dfrac{dH}{dt} = 0$. Substituting these conditions into system (2.1) yields the following relationships among the parameters: $I_b = aG_f + bH_f, \quad H_b = cH_f - dG_f$. To express the system in terms of deviations from the fasting state, we define new variables $g(t) = G(t) - G_f, \quad h(t) = H(t) - H_f$, which represent fluctuations in glucose and insulin levels relative to their fasting values. Substituting these into the original system (2.1) and simplifying using the steady-state conditions, we obtain the following reduced system of differential equations:

$$\begin{cases} \dfrac{dg}{dt} = -ag - bh + I(t), \\ \dfrac{dh}{dt} = -ch + dg. \end{cases} \tag{2.2}$$

2.2.2 Model Reduction to a Second-Order ODE and Its Analytical Solution

In this section, we reduce the system of two first-order ordinary differential equations (2.2) to a single second-order equation. Starting with the first equation in the system, we differentiate both sides with respect to time to obtain:

$$\frac{d^2g}{dt^2} = -a\frac{dg}{dt} - b\frac{dh}{dt} + \frac{dI}{dt}.$$

To eliminate the variable h, we use the original system (2.2). Multiply the first equation by $-c$ and the second by b:

$$-c\frac{dg}{dt} = acg + bch - cI,$$
$$b\frac{dh}{dt} = -bch + bdg.$$

Adding these equations and isolating $b\frac{dh}{dt}$ yields:

$$b\frac{dh}{dt} = c\frac{dg}{dt} + (ac + bd)g - cI.$$

Substituting this expression into the second derivative of g, we obtain:

$$\frac{d^2g}{dt^2} + (a + c)\frac{dg}{dt} + (ac + bd)g = \frac{dI}{dt} + cI(t).$$

Introducing parameters $\alpha = \frac{a+c}{2}$, $\omega_0^2 = ac + bd$, and $X(t) = \frac{dI}{dt} + cI(t)$, we rewrite the model as:

$$\frac{d^2g}{dt^2} + 2\alpha\frac{dg}{dt} + \omega_0^2 g = X(t), \tag{2.3}$$

which has the standard form of the damped, forced harmonic oscillator studied in classical ODEs. Here, ω_0 represents the natural frequency and α is the damping coefficient. Thus, the dynamics of blood glucose concentration can be described using a second-order linear differential equation with two effective parameters, replacing the original four.

Next, we solve the model (2.3) for $g(t)$. The right-hand side function $X(t)$ can be approximated by a Dirac delta function, $X(t) \approx F\delta(t)$, where F is a constant. This approximation is reasonable since $X(t)$ depends solely on the external glucose input $I(t)$, which we assume to occur instantaneously, such as a single meal ingested at a specific moment. The Dirac delta function captures this behavior as a mathematical idealization of a very short, intense impulse. With this approximation, the model reduces to the following initial value problem:

$$\frac{d^2g}{dt^2} + 2\alpha\frac{dg}{dt} + \omega_0^2 g = F\delta(t), \quad g(0) = 0, \quad \frac{dg}{dt}(0) = 0. \tag{2.4}$$

In the equation above, the initial conditions indicate that, at the initial time (the moment just before food intake), the blood glucose concentration is at equilibrium, and there is no initial change in glucose levels.

Applying the Laplace transform [6] to both sides of (2.4) gives

$$\left[s^2\mathcal{L}\{g(t)\} - sg(0) - \left.\frac{dg}{dt}\right|_{t=0}\right] + 2\alpha\left[s\mathcal{L}\{g(t)\} - g(0)\right] + \omega_0^2\mathcal{L}\{g(t)\} = F.$$

Substituting the initial conditions we obtain:

$$s^2\mathcal{L}\{g(t)\} + 2\alpha s\mathcal{L}\{g(t)\} + \omega_0^2\mathcal{L}\{g(t)\} = F.$$

Solving for $\mathcal{L}\{g(t)\}$, we get

$$\mathcal{L}\{g(t)\} = \frac{F}{s^2 + 2\alpha s + \omega_0^2}.$$

To find the solution $g(t)$, we apply the inverse Laplace transform:

$$g(t) = \mathcal{L}^{-1}\left\{\frac{F}{s^2 + 2\alpha s + \omega_0^2}\right\} = \mathcal{L}^{-1}\left\{\frac{F}{(s+\alpha)^2 + \omega^2}\right\} = \frac{F}{\omega}e^{-\alpha t}\sin(\omega t), \tag{2.5}$$

where $\omega^2 = \omega_0^2 - \alpha^2$.

Therefore, the glucose concentration $G(t) = g(t) + G_f$ over time is given by:

$$G(t) = G_f + \frac{F}{\omega}e^{-\alpha t}\sin(\omega t). \tag{2.6}$$

To derive the solution for $H(t) = h(t) + H_f$, we substitute the expression (2.5) for $g(t)$ into the second equation in system (2.2). This yields a nonhomogeneous linear differential equation for $h(t)$ with known forcing:

$$\frac{dh}{dt} + ch = d\,g(t),$$

where $g(t)$ is defined by (2.5).

This equation is a first-order linear ODE and can be solved using standard techniques, such as the integrating factor method. Applying this method and incorporating the initial condition $h(0) = 0$, we arrive at the explicit solution:

$$H(t) = H_f + \frac{dF}{\omega\left(\omega^2 + (\alpha - c)^2\right)}\left[\omega\left(e^{-ct} - e^{-\alpha t}\cos(\omega t)\right) - (\alpha - c)e^{-\alpha t}\sin(\omega t)\right]. \tag{2.7}$$

The derived expressions (2.6) and (2.7) describe the time-dependent behavior of glucose concentration and insulin level.

2.2.3 Fitting the Model to Data from Healthy and Diabetic Patients

Having a closed-form solution to the glucose-insulin model allows us to fit the model directly to real observational data. In what follows, we fit the solution for glucose concentration (2.6) to measured blood glucose levels following an oral glucose load in both healthy and diabetic individuals. This fitting yields parameter values that can later be used to explore and evaluate potential treatment strategies for improving diabetes management.

The blood glucose function $G(t)$ from (2.6) is fitted to the data shown in Table 2.1, which includes measurements from a non-diabetic and a diabetic individual [4]. Note that blood glucose concentrations are typically reported in milligrams per deciliter (mg/dL).

Figure 2.3 presents the measured blood glucose concentration data for a non-diabetic individual in response to a glucose tolerance test, along with the corresponding fitted model. Figure 2.4 shows the same for a diabetic individual. The model fitting was performed in MATLAB [7] using the least squares method. The source code is available on GitHub [8].

Table 2.1 Observed blood glucose data for non-diabetic and diabetic individuals

	Non-diabetic Individual		Diabetic Individual
t, min	G, mg/dL	t, min	G, mg/dL
0	75	0	85
30	95	30	160
60	92	60	196
90	78	120	130
120	63	180	85
150	67	240	60
180	74		

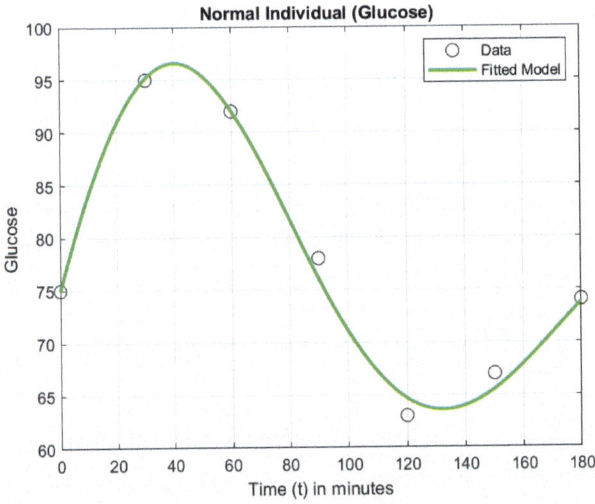

Fig. 2.3 Fitted model $G(t)$ to measured blood glucose levels of a non-diabetic individual

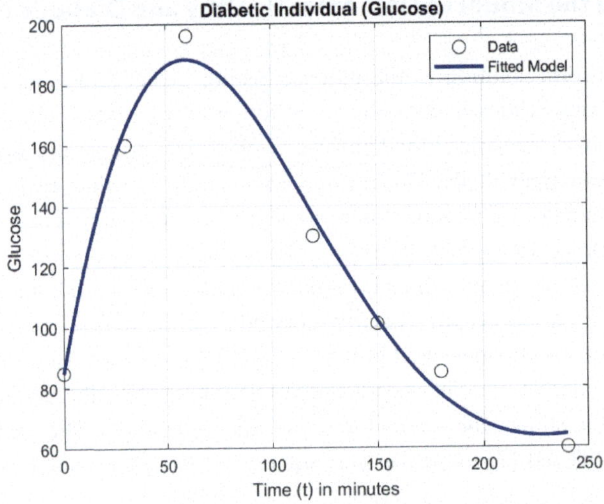

Fig. 2.4 Fitted model $G(t)$ to measured blood glucose levels of a diabetic individual

Table 2.2 contains parameters extracted from the fitting for both non-diabetic and diabetic models.

Figure 2.5 shows the model solutions extended over a longer time period for both non-diabetic and diabetic individuals. Notice that in the non-diabetic case, blood glucose levels return to baseline in approximately $T = 180$ minutes (3 hours), while for the diabetic individual, this return takes about $T = 300$ minutes (5 hours). Thus, it takes nearly twice as long for the diabetic individual's glucose level to return to the fasting baseline. Another noticeable difference appears in the maximum glucose concentration: the non-diabetic

Table 2.2 Model parameters extracted by fitting the glucose function (2.6) to data from a non-diabetic and a diabetic individual

Model parameters	G_f, mg/dL	α, $1/min$	ω, rad/min	F $(mg \cdot rad)/(dL \cdot min)$	$T = \frac{2\pi}{\omega_0}$, min
Non-diabetic	75	0.0070	0.0341	0.996	180
Diabetic	85	0.0094	0.0185	3.740	300

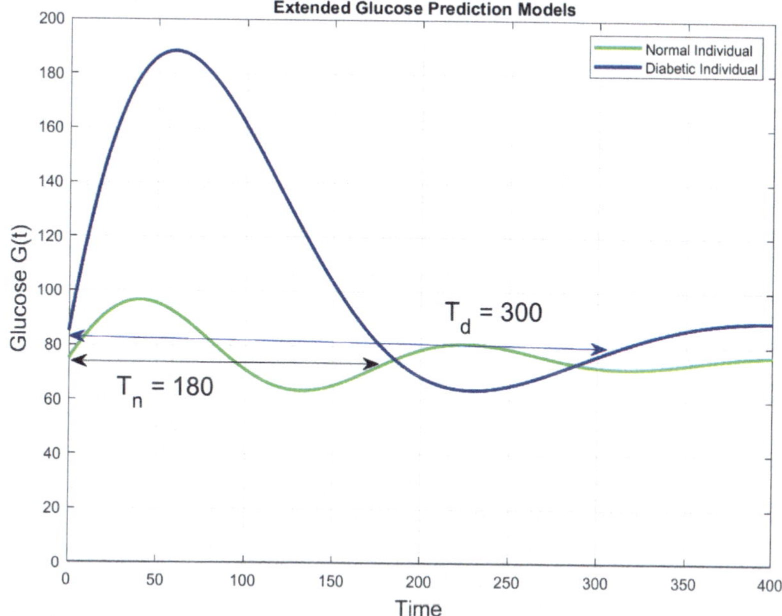

Fig. 2.5 Extended glucose models for a non-diabetic and a diabetic individual. The period for the healthy individual is $T_n = 180$ minutes (3 hours), while for the diabetic individual it is $T_d = 300$ minutes (5 hours)

individual reaches a maximum of about 100 mg/dL, while the diabetic individual peaks at around 190 mg/dL. Both recovery time and maximum glucose concentration can serve as diagnostic indicators to differentiate between diabetic and non-diabetic individuals [4, 5].

Our ultimate objective is to apply the derived glucose-insulin model to evaluate various diabetes management strategies. To do so, we must determine specific values for the original model parameters in the system (2.2). Estimating the four parameters,a, b, c, and d requires four independent conditions. From fitting the glucose function $G(t)$ to the measured glucose data, we obtained estimates for the parameters F, α, and ω, as shown in Table 2.2. These are related to the original parameters through the identities $2\alpha = a+c$ and $\omega^2 - \alpha^2 = ac + bd$, but these relationships alone do not suffice to uniquely determine the four parameters. Therefore, we turn to the explicit expression for the insulin function $H(t)$ given in (2.7) and fit it to the corresponding insulin data for the same diabetic individual

Table 2.3 Observed blood glucose and insulin levels for a diabetic individual

t, min	$G, mg/dL$	$H, \mu U/mL$
0	85	20
30	160	60
60	196	132
120	130	200
180	85	76
240	60	30

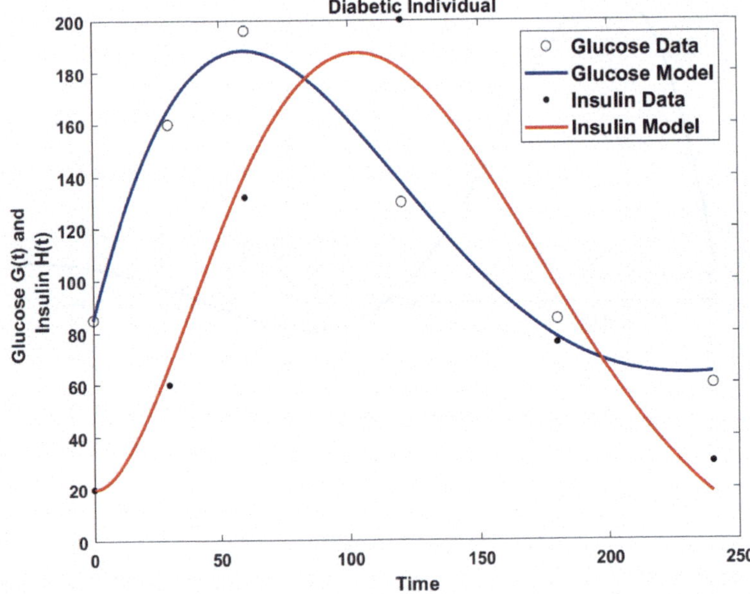

Fig. 2.6 Glucose and insulin models $G(t)$ and $H(t)$ fitted to observed data from a diabetic individual

[4]. The datasets for both glucose and insulin levels are presented in Table 2.3. Note that insulin levels are typically reported in microunits per milliliter ($\mu U/mL$).

Figure 2.6 illustrates the glucose (blue) and insulin (red) models, each fitted to the corresponding data in Table 2.3. The parameter values for α, ω, and F are the same as those obtained earlier, as presented in Table 2.2. By fitting the insulin function $H(t)$ from (2.6) to the insulin data from Table 2.3, we determined the values for $c = 0.0179$ (min^{-1}) and $d = 0.0414$ ($\frac{\mu\text{U}}{\text{mL}} \cdot \frac{\text{dL}}{\text{mg}} \cdot \frac{1}{min}$). Using the relationships between the two parameter sets, we then obtained $a = 0.001$ (min^{-1}) and $b = 0.010$ ($\frac{\text{mg}}{\text{dL}} \cdot \frac{\text{mL}}{\mu\text{U}} \cdot \frac{1}{min}$). These parameter values can now be used in the model to test different treatment regimens and determine which one yields the best outcome.

2.2.4 Application of the Model to Diabetes Treatment Strategies

Next, we apply the derived model to simulate a patient's blood glucose concentration under three distinct scenarios, each based on the type of insulin administration and the mathematical representation of food intake. Our objective is to evaluate the impact of timing in insulin delivery and to test the hypothesis that insulin administration before food intake leads to better glucose control. In our simulations, food intake is modeled using a Dirac delta function appropriate when meal is consumed over a very short period relative to the time scale of glucose dynamics or a Heaviside step function, which represents a longer and sustained intake. Similarly, insulin delivery is represented as either a delta function, corresponding to a single injection, or as a Heaviside function, corresponding to a continuous infusion over a controlled period, such as via an insulin pump.

Insulin and Food Intake Modeled as Delta Function Inputs

We begin by applying the derived model to simulate insulin treatment administered via injection for a diabetic individual. Let both the insulin injection and the food intake modeled as instantaneous events, represented mathematically by Dirac delta functions. Under these assumptions, the glucose-insulin dynamics are governed by the following initial value problem:

$$\begin{cases} \dfrac{dg}{dt} = -ag - bh + I(t), \\ \dfrac{dh}{dt} = -ch + dg + K(t), \\ g(0) = 0, \quad h(0) = 0, \end{cases} \tag{2.8}$$

where $I(t) = I_0\delta(t - t_1)$ represents the effect of food intake, and $K(t) = K_0\delta(t - t_2)$ models the insulin injection, as illustrated in Fig. 2.7. Since $g(t)$ and $h(t)$ represent deviations from the fasting levels, we assume zero initial conditions.

We explore three scenarios: insulin and food are administered simultaneously ($t_1 = t_2$); food is consumed before insulin is injected ($t_1 < t_2$); and insulin is injected prior to food

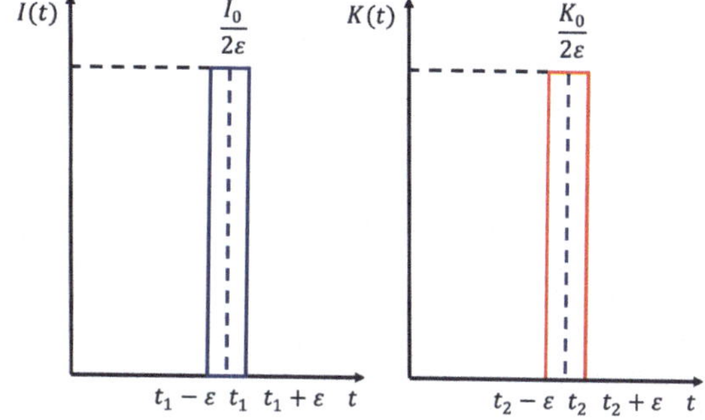

Fig. 2.7 Visualization of food intake at time t_1 and insulin injection at time t_2, both modeled as Dirac delta functions

intake ($t_1 > t_2$). The goal is to evaluate the glucose response in each case and to assess whether administering insulin before eating improves glucose regulation.

The model parameters used are:

$$a = 0.001\,\text{min}^{-1}, \quad b = 0.010\,\frac{\text{mg}}{\text{dL}} \cdot \frac{\text{mL}}{\mu\text{U}} \cdot \frac{1}{\text{min}}, \quad c = 0.018\,\text{min}^{-1},$$

$$d = 0.041\,\frac{\mu\text{U}}{\text{mL}} \cdot \frac{\text{dL}}{\text{mg}} \cdot \frac{1}{\text{min}}.$$

Let us solve the model (2.8) using Laplace transform [6].

$$\begin{cases} \mathcal{L}\{\frac{dg}{dt}\} + a\mathcal{L}\{g\} & = -b\mathcal{L}\{h\} + I_0\mathcal{L}\{\delta(t-t_1)\} \\ \mathcal{L}\{\frac{dh}{dt}\} + c\mathcal{L}\{h\} & = d\mathcal{L}\{g\} + K_0\mathcal{L}\{\delta(t-t_2)\} \end{cases}$$

Using the properties of Laplace transform and applying them to the derivatives and delta functions reduces this system to:

$$\begin{cases} \mathcal{L}\{g\}(s+a) & = -b\mathcal{L}\{h\} + I_0 e^{-st_1} \\ \mathcal{L}\{h\}(s+c) & = d\mathcal{L}\{g\} + K_0 e^{-st_2}. \end{cases}$$

Using that

$$(s+a) + \frac{bd}{(s+c)} = \frac{(s+\alpha)^2 + \omega^2}{(s+c)},$$

we can solve for $\mathcal{L}\{g\}$ and $\mathcal{L}\{h\}$ to get the following:

$$\begin{aligned} \mathcal{L}\{g\} &= \frac{(s+c)I_0 e^{-st_1} - bK_0 e^{-st_2}}{(s+\alpha)^2 + \omega^2}, \\ \mathcal{L}\{h\} &= \frac{d(s+c)I_0 e^{-st_1} + K_0 e^{-st_2}((s+\alpha)^2 + \omega^2) - bd}{(s+c)((s+\alpha)^2 + \omega^2)} \end{aligned} \tag{2.9}$$

To determine the time-dependent solutions for glucose and insulin, $g(t)$ and $h(t)$, we apply the inverse Laplace transform to the system using MATLAB for the numerical computation. The code for the inverse Laplace transform is available in [8]. We simulated three insulin treatment scenarios: insulin administered simultaneously with food intake, 30 minutes after food intake, and 30 minutes before food intake. The results, shown in Fig. 2.8, indicate that administering insulin 30 minutes prior to food intake yields the most favorable outcome: both glucose and insulin responses (shown in orange) exhibit the smallest peak amplitudes and fastest return to baseline. The analytical derivation of $g(t)$ and $h(t)$ is left as an exercise.

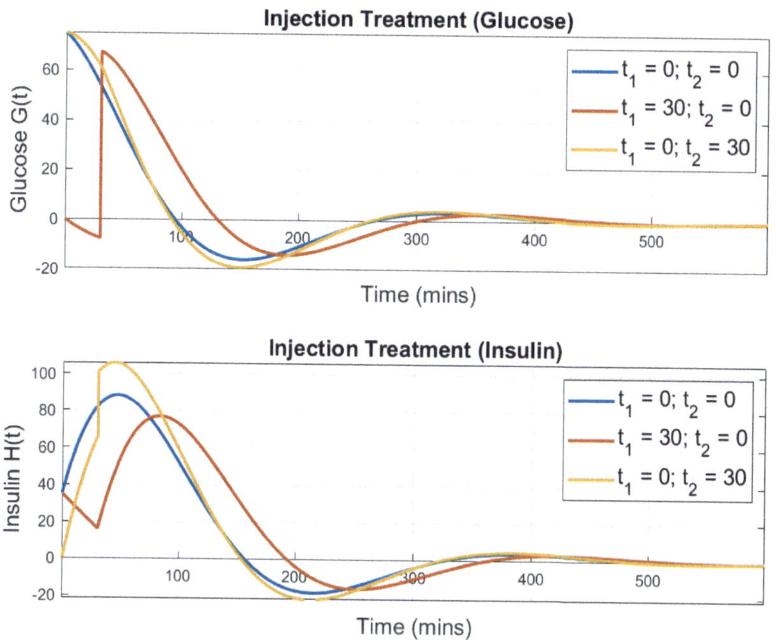

Fig. 2.8 Injection treatment scenarios: (**a**) insulin and food intake occur simultaneously at $t_1 = t_2$ (blue curves); (**b**) insulin is administered at $t_2 = 0$, followed by food intake at $t_1 = 30$ minutes (orange curves); (**c**) food intake occurs at $t_1 = 0$, followed by insulin injection at $t_2 = 30$ minutes (yellow curves)

These findings indicate that the optimal treatment for a diabetic individual occurs when insulin is administered approximately 30–40 minutes prior to food intake. This highlights the critical role of timing in diabetes management and demonstrates how mathematical modeling can guide more informed treatment strategies.

We explore four scenarios of early administration of insulin injections to determine the most effective timing relative to food intake. Specifically, we consider insulin administered 15, 30, 45, or 60 minutes prior to eating. The corresponding glucose and insulin responses for each case are shown in Fig. 2.9. The results suggest that administering insulin 30–45 minutes before food intake provides the most favorable outcome. Notably, when insulin is given more than 45 minutes in advance, the amplitude of the glucose response begins to rise again, indicating reduced treatment effectiveness.

Modeling Food Intake with a Step Function and Insulin Treatment with a Delta Function

In the previous case, food intake was represented by a Dirac delta function to model an instantaneous glucose input. Here, we assume that food is consumed over a relatively short period (say, on the order of 10 minutes),which is still brief compared with the overall time scale of glucose regulation. This scenario can be modeled using a step function $I(t) = I_0(U_{\tau_1}(t) - U_{\tau_2}(t))$, where τ_1 and τ_2 represent the start and end times of food intake

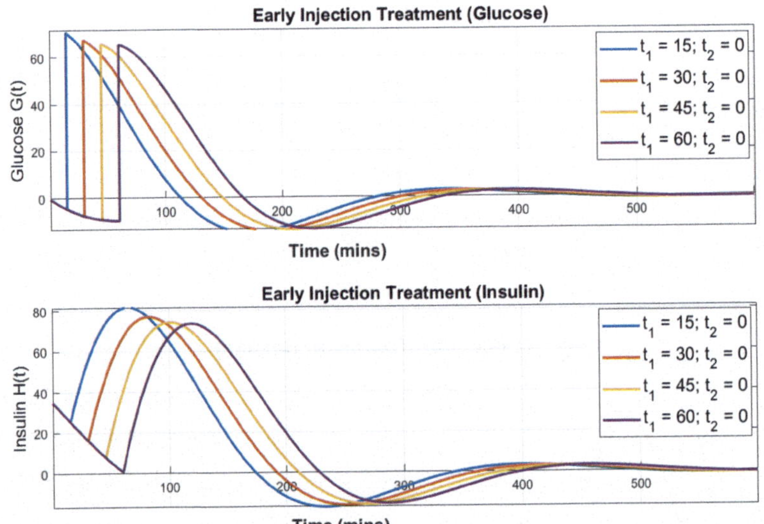

Fig. 2.9 Injection treatment scenarios with insulin administered prior to food intake: (**a**) 15 minutes before ($t_2 = 0, t_1 = 15$), shown in blue; (**b**) 30 minutes before ($t_2 = 0, t_1 = 30$), shown in orange; (**c**) 45 minutes before ($t_2 = 0, t_1 = 45$), shown in yellow; and (**d**) 60 minutes before ($t_2 = 0, t_1 = 60$), shown in purple

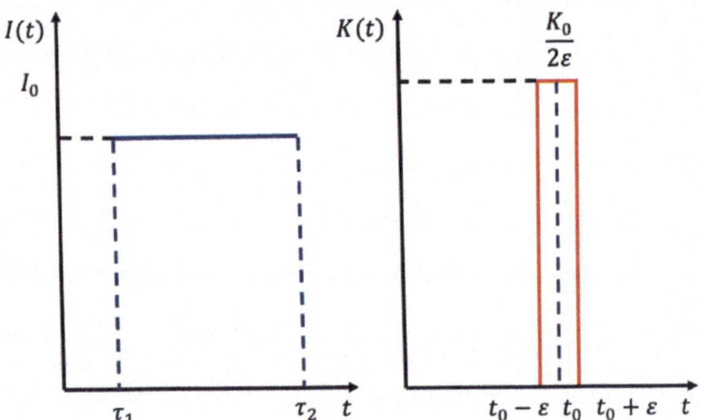

Fig. 2.10 Visualization of food intake modeled as a step function and insulin injection represented by a Dirac delta function

respectively. Here $U_\tau(t)$ is the unit step function shifted to time τ. The insulin treatment is still administered as an injection and modeled by a Dirac delta function, $K(t) = K_0\delta(t - t_0)$. These functions are illustrated in Fig. 2.10. The height and duration of the step function are chosen so that its total area matches the area under the delta function shown in Fig. 2.9, ensuring the same total amount of glucose intake is represented.

Applying Laplace transform to the initial value problem (2.8) we get

$$\begin{cases} \mathcal{L}\{g\}(s+a) &= -b\mathcal{L}\{h\} + I_0\left(\dfrac{e^{-s\tau_1}}{s} - \dfrac{e^{-s\tau_2}}{s}\right) \\ \mathcal{L}\{h\}(s+c) &= d\mathcal{L}\{g\} + K_0 e^{-st_0}. \end{cases} \tag{2.10}$$

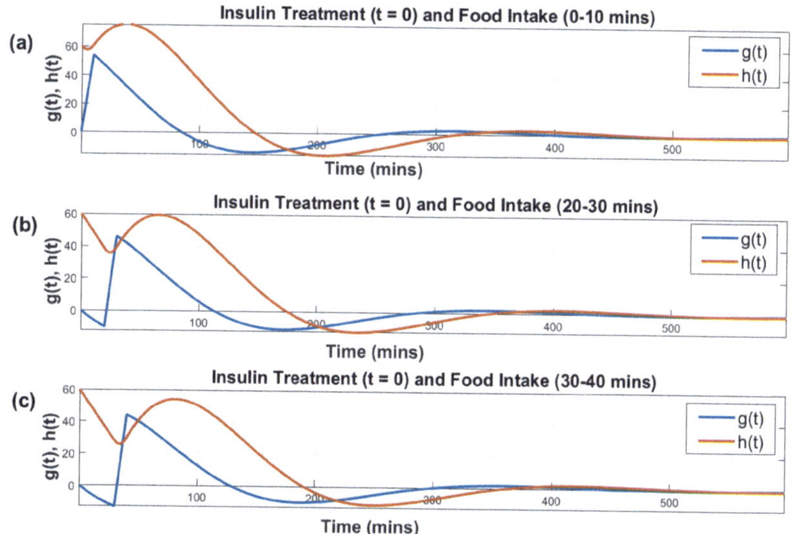

Fig. 2.11 Glucose (blue) and insulin (red) concentration curves when food is consumed during the intervals: (**a**) 0–10 minutes, (**b**) 20–30 minutes, and (**c**) 30–40 minutes after insulin injection

One can solve for $\mathcal{L}\{g\}$ and $\mathcal{L}\{h\}$ to get the following:

$$\begin{aligned}\mathcal{L}\{g\} &= \frac{(s+c)}{(s+\alpha)^2+\omega^2}\left[I_0\left(\frac{e^{-s\tau_1}-e^{-s\tau_2}}{s}\right) - bK_0\frac{e^{-st_0}}{(s+c)}\right] \\ \mathcal{L}\{h\} &= \frac{d}{(s+c)}\mathcal{L}\{g\} + \frac{K_0}{(s+c)}e^{-st_0}.\end{aligned} \tag{2.11}$$

Next, we use MATLAB to compute the inverse Laplace transform and obtain the functions $g(t)$ and $h(t)$; the corresponding code is available in [8]. Figure 2.11 presents numerical simulations of glucose and insulin levels over time. The glucose curve, shown in blue, exhibits the lowest initial peak and the shortest duration when food intake occurs 30–40 minutes after insulin administration. These results are consistent with our earlier findings when both food intake and insulin treatment were modeled using Dirac delta functions.

Modeling Food Intake and Insulin Treatment Using Step Functions

Finally, we assume that food is consumed over a short period of time and the insulin treatment is administered as continuous infusion over a fixed duration, for example, via an insulin pump.

Both the food intake and the insulin treatment are modeled as step functions: $I(t) = I_0(U_{\tau_1}(t) - U_{\tau_2}(t))$ and $K(t) = K_0(U_{\tau_3}(t) - U_{\tau_4}(t))$, where the time intervals $[\tau_1, \tau_2]$ and $[\tau_3, \tau_4]$ represent the periods for food intake and insulin administration, respectively, as illustrated in Fig. 2.12. Note that the height and duration of the step functions are chosen

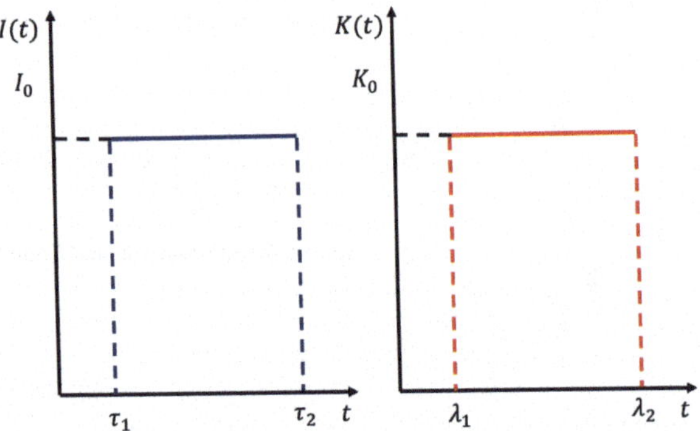

Fig. 2.12 Visualization of both food intake and insulin treatment modeled as step functions, where food is consumed over the interval $[\tau_1, \tau_2]$ and the treatment, administered as a pill, dissolves over the interval $[\tau_3, \tau_4]$

to ensure that the total amounts of food intake and insulin administration match those represented by the delta functions shown in Fig. 2.7.

Applying Laplace transform to the initial value problem (2.8) we get the system of two equations below

$$\begin{cases} \mathcal{L}\{\frac{dg}{dt}\} + a\mathcal{L}\{g\} & = -b\mathcal{L}\{h\} + I_0\mathcal{L}\{U_{\tau_1}(t) - U_{\tau_2}(t)\} \\ \mathcal{L}\{\frac{dh}{dt}\} + c\mathcal{L}\{h\} & = d\mathcal{L}\{g\} + K_0\mathcal{L}\{(U_{\tau_3}(t) - U_{\tau_4}(t))\}. \end{cases}$$

Applying Laplace transform properties to the derivatives and step functions, reduces this system to

$$\begin{cases} \mathcal{L}\{g\}(s+a) & = -b\mathcal{L}\{h\} + I_0\left(\frac{e^{-s\tau_1}}{s} - \frac{e^{-s\tau_2}}{s}\right) \\ \mathcal{L}\{h\}(s+c) & = d\mathcal{L}\{g\} + K_0\left(\frac{e^{-s\tau_3}}{s} - \frac{e^{-s\tau_4}}{s}\right). \end{cases} \tag{2.12}$$

We can solve for $\mathcal{L}\{g\}$ and $\mathcal{L}\{h\}$ to get the following:

$$\begin{aligned} \mathcal{L}\{g\} &= \frac{s+c}{(s+\alpha)^2+\omega^2}\left[I_0\left(\frac{e^{-s\tau_1}-e^{-s\tau_2}}{s}\right) - \frac{bK_0}{(s+c)}\left(\frac{e^{-s\tau_3}-e^{-s\tau_4}}{s}\right)\right]. \\ \mathcal{L}\{h\} &= \frac{d}{(s+c)}\mathcal{L}\{g\} + \frac{K_0}{(s+c)}\left(\frac{e^{-s\tau_3}}{s} - \frac{e^{-s\tau_4}}{s}\right). \end{aligned} \tag{2.13}$$

As in previous cases, we use the inverse Laplace transform in MATLAB to solve for g and h. The resulting glucose and insulin curves are presented in Fig. 2.13. In panel (a), where food and insulin infusion occur simultaneously, and in panel (c), where food is consumed first followed by the infusion, glucose reaches an initial peak of 50 mg/dL, and insulin also peaks at 100 μU/mL or higher. In contrast, when the insulin infusion is administered before the food, as shown in panel (b), the treatment is optimal: the initial glucose peak is significantly lower, around 30 mg/dL, and insulin levels are more moderate.

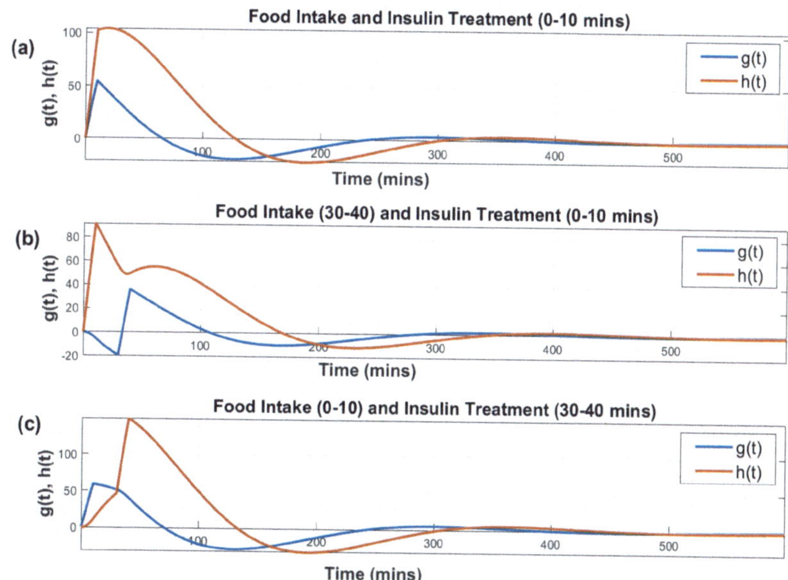

Fig. 2.13 Glucose (in blue) and insulin (in red) curves for: (**a**) both food intake and insulin treatment applied between 0 and 10 minutes; (**b**) insulin treatment applied between 0 and 10 minutes, with food intake occurring between 30 and 40 minutes; and (**c**) food intake first between 0 and 10 minutes, followed by insulin treatment between 30 and 40 minutes

In this section, we presented a simplified mathematical model of the glucose-insulin regulatory system and used it to explore scenarios relevant to the management of type 2 diabetes. As well documented in the medical literature and as our simulation results illustrate, insulin treatment can cause blood glucose levels to fall below the baseline, which can lead to adverse symptoms such as dizziness, weakness, or fatigue. Several treatment strategies were examined by modeling insulin injections as Dirac delta functions and continuous infusions as Heaviside step functions, with a focus on the timing and form of intervention. The analysis indicates that initiating treatment approximately 30–40 minutes before food intake results in more effective regulation of blood glucose levels.

Although these simulations help illustrate general trends, we have so far chosen insulin dosage and timing by trial and error. In the next section we apply optimal control theory to determine mathematically optimal treatment while accounting for practical limitations such as maximal insulin dosing and timing restrictions.

2.3 Optimal Control for Diabetes Management

In this section, we introduce optimal control theory [9–11] in the context of blood glucose regulation. Our objective is to determine a control function $u(t)$,the insulin infusion rate over time, that keeps blood glucose levels close to a desired constant target while minimizing the total amount of insulin administered.

In Sect. 2.2, external insulin inputs were modeled using the function $K(t)$, either as a Dirac delta function (representing a bolus injection) or a Heaviside step function (for infusion over a short interval). In contrast, the control input $u(t)$ introduced here is a continuous function of time. It represents an adjustable infusion rate that will be determined by solving an optimal control problem.

2.3.1 Model Setup and Control Objectives

To formalize the problem, we introduce a cost functional, a rule that assigns a single numerical value to each possible control function $u(t)$ that represents the general "cost" of that control in achieving our objectives. The cost functional reflects two competing objectives: maintaining blood glucose levels close to a fixed target l and maintaining the amount of insulin administered. These objectives are quantified by the squared deviation $(g(t) - l)^2$ and the squared control $u(t)^2$, respectively. A positive weighting parameter $\rho > 0$ balances the relative importance of these terms, yielding the cost functional:

$$\mathcal{J}(u) = \int_0^{\tau} ((g(t) - l)^2 + \rho u(t)^2) dt, \tag{2.14}$$

where $\tau > 0$ is the duration of time control period.

The integral form of the cost functional reflects the fact that regulation quality is evaluated over the entire interval $[0, \tau]$. Both the squared deviation of glucose from the target and the insulin effort are accumulated in time, so that poor control at any stage contributes to the total cost.

Recall the glucose-insulin regulation model introduced in the previous section, specifically the system of differential equations given in (2.2). For the purposes of optimal control, we rewrite this system as follows:

$$\begin{cases} \dfrac{dg}{dt} = -ag - bh, \\ \dfrac{dh}{dt} = -ch + dg + u. \end{cases} \tag{2.15}$$

In formulating system (2.15), we make the following assumptions:

- There is no external glucose input i.e. $I(t) = 0$;
- Control $u(t)$ is applied over the interval $[0, \tau]$, where we initially assume that $\tau \to \infty$. The assumption that $\tau \to \infty$ is known as the "infinite horizon assumption" meaning that control is applied over an unbounded time interval. The infinite horizon formulation

is often used because it simplifies mathematical analysis and provides insight into the long-term behavior of the system. This assumption will be relaxed in Sect. 2.3.4, where we consider optimal finite horizon control, that is, optimization over a fixed and limited time period.

Our goal is to find the control function $u(t)$ that minimizes the cost functional $\mathcal{J}(u)$ subject to the dynamical system (2.15). This is known as the Linear Quadratic Regulator (LQR) problem, since the system is linear in the state variables g, h, and the control u, and the cost functional is quadratic in both the state and control.

To proceed, we first review the general formulation and solution of the infinite horizon LQR problem. We then apply this framework to the glucose-insulin regulation model.

2.3.2 Overview of the Linear Quadratic Regulator (LQR) Framework

We begin with a brief review of the general structure of the Linear Quadratic Regulator (LQR) problem [9], a fundamental framework in optimal control theory. The LQR problem involves finding a control input $\mathbf{u}(t)$ that regulates the state $\mathbf{x}(t)$ of a linear dynamical system to minimize a quadratic cost functional $\mathcal{J}(\mathbf{u})$. The LQR framework is well established in engineering and control system analysis; here, we present its general formulation before adapting it to the glucose–insulin regulation problem in Sect. 2.3.3.

Let us consider a linear dynamical system with state equations:

$$\frac{d\mathbf{x}}{dt} = A(t)\mathbf{x} + B(t)\mathbf{u}, \quad \mathbf{x}(0) = \mathbf{x}_0, \tag{2.16}$$

where $\mathbf{x}(t) \in \mathbb{R}^n$ is the vector of state variables, and $\mathbf{u}(t) \in \mathbb{R}^m$ is the control vector. The matrices $A \in \mathbb{R}^{n\times n}$ and $B \in \mathbb{R}^{n\times m}$ may vary with time, although many applications use constant matrices.

The goal is to minimize the cost functional:

$$\mathcal{J}(\mathbf{u}) = \frac{1}{2}\int_0^{\tau} \left[\mathbf{e}^T Q(t)\mathbf{e} + \mathbf{u}^T R(t)\mathbf{u}\right] dt, \tag{2.17}$$

where $\mathbf{e}(t) = \mathbf{x}(t) - \boldsymbol{l}$ denotes the deviation from a constant target level $\boldsymbol{l} \in \mathbb{R}^n$. The matrix $Q(t) \in \mathbb{R}^{n\times n}$ is symmetric and positive semidefinite, meaning $\mathbf{w}^T Q\mathbf{w} \geq 0$ for all $\mathbf{w} \in \mathbb{R}^n$, and it weights the importance of keeping the state close to the target. The matrix $R(t) \in \mathbb{R}^{m\times m}$ is symmetric and positive definite, meaning $\mathbf{v}^T R\mathbf{v} > 0$ for all non-zero $\mathbf{v} \in \mathbb{R}^m$, and it penalizes large control inputs. Both $Q(t)$ and $R(t)$ can be continuous functions of time. The factor $\frac{1}{2}$ is a standard convention that simplifies derivatives without changing the optimal solution.

To solve this problem, we introduce a vector of Lagrange multipliers (also called the adjoint vector) $\boldsymbol{\lambda}(t) \in \mathbb{R}^n$, and define the Hamiltonian function:

$$\mathcal{H}(\mathbf{x}, \mathbf{u}, \boldsymbol{\lambda}, t) = \frac{1}{2}\mathbf{e}^T Q(t)\mathbf{e} + \frac{1}{2}\mathbf{u}^T R(t)\mathbf{u} + \boldsymbol{\lambda}^T A(t)\mathbf{x} + \boldsymbol{\lambda}^T B(t)\mathbf{u}. \tag{2.18}$$

The Hamiltonian combines the instantaneous cost and the system dynamics. Lagrange multipliers represent how sensitive the total cost is to changes in the state.

In terms of the Hamiltonian, the cost functional can be rewritten as:

$$\mathcal{J} = \frac{1}{2}\int_0^\tau \left[\mathcal{H} - \boldsymbol{\lambda}^T \frac{d\mathbf{x}}{dt}\right] dt. \tag{2.19}$$

Pontryagin's Maximum Principle [10] provides conditions that a control must satisfy to minimize the total cost functional $\mathcal{J}(\mathbf{u})$; these are known as the necessary conditions for optimality.

The necessary conditions for optimality are:

The state equation:

$$\frac{d\mathbf{x}}{dt} = \frac{\partial \mathcal{H}}{\partial \boldsymbol{\lambda}}, \qquad \mathbf{x}(0) = \mathbf{x}_0, \tag{2.20}$$

which simply restates the system dynamics.

The costate (adjoint) equation:

$$\frac{d\boldsymbol{\lambda}}{dt} = -\frac{\partial \mathcal{H}}{\partial \mathbf{x}}, \tag{2.21}$$

describing how the sensitivity of the cost evolves backward in time.

The optimality condition:

$$\frac{\partial \mathcal{H}}{\partial \mathbf{u}} = 0, \tag{2.22}$$

which gives the control input that minimizes the cost at each instant.

The transversality condition:

$$\boldsymbol{\lambda}(\tau) = 0. \tag{2.23}$$

ensuring the cost stops at the end of the control period.

The optimality condition (2.22) can be used to obtain

$$\mathbf{u} = -R^{-1}B^T\boldsymbol{\lambda} \tag{2.24}$$

Using the adjoint equation (2.21) together with the transversality condition (2.23), one can derive:

$$\frac{d\boldsymbol{\lambda}}{dt} = -Q\mathbf{x} - A^T\boldsymbol{\lambda} + Ql, \quad \boldsymbol{\lambda}(\tau) = \mathbf{0} \tag{2.25}$$

For linear systems with quadratic costs, the adjoint can be expressed as a linear function of the state vector:

$$\boldsymbol{\lambda}(t) = P(t)\mathbf{x} + \boldsymbol{\mu}(t), \tag{2.26}$$

where $P(t) \in \mathbb{R}^{n\times n}$ is a symmetric positive semi-definite matrix and $\boldsymbol{\mu}(t) \in \mathbb{R}^n$ is a column vector. This representation simplifies the solution by reducing the costate equation to equations for $P(t)$ and $\boldsymbol{\mu}(t)$.

Substituting (2.26) into (2.25) leads to the matrix Riccati differential equation:

$$\frac{dP}{dt} = -PA - A^TP + PBR^{-1}B^TP - Q, \quad P(\tau) = 0 \tag{2.27}$$

which is integrated backward in time from $t = \tau$ to $t = 0$. Conceptually, this is similar to solving a standard ODE, but for the elements of a matrix and in reverse time.

The vector $\boldsymbol{\mu}(t)$ satisfies:

$$\frac{d\boldsymbol{\mu}}{dt} = (PBR^{-1}B^T - A^T)\boldsymbol{\mu} + Ql, \quad \boldsymbol{\mu}(\tau) = 0 \tag{2.28}$$

and is also integrated backward in time.

The solution of the LQR optimal control problem can now be summarized as a step-by-step procedure. Each step may be carried out analytically or numerically, depending on the system parameters and the complexity of the functions involved:

Step 1: Solve the matrix Riccati equation (2.27) backward in time to find $P(t)$.

Step 2: Solve the Eq. (2.28) backward in time to obtain $\boldsymbol{\mu}(t)$.

Step 3: Given $P(t)$ and $\boldsymbol{\mu}(t)$, use (2.26) to construct the adjoint variable $\boldsymbol{\lambda}$. Substitute $\boldsymbol{\lambda}$ into the Eq. (2.24) to obtain the optimal control $\mathbf{u}(t) = -R^{-1}B^T(P(t)\mathbf{x} + \boldsymbol{\mu})$.

Step 4: Substitute the control into the state equation to form the system:

$$\frac{d\mathbf{x}}{dt} = \left(A - BR^{-1}B^TP(t)\right)\mathbf{x}(t) - BR^{-1}B^T\boldsymbol{\mu}(t), \quad \mathbf{x}(0) = \mathbf{x}_0.$$

Solve this system to find $\mathbf{x}(t)$.

Step 5: Once $\mathbf{x}(t)$ is known, we substitute it into the feedback control formula from Step 3 to obtain the explicit time-dependent solution for the optimal control $\mathbf{u}(t)$.

Having outlined the general LQR solution procedure, we now apply it to the glucose–insulin regulation model (2.15).

2.3.3 Application of LQR to Blood Glucose Regulation

The general LQR framework developed above is now applied to the glucose–insulin model to determine the optimal insulin delivery strategy.

The system (2.15) fits the form of the general linear system (2.16) with

$$A = \begin{bmatrix} -a & -b \\ d & -c \end{bmatrix}, \quad B = \begin{bmatrix} 0 \\ 1 \end{bmatrix}, \quad \mathbf{x} = \begin{bmatrix} g \\ h \end{bmatrix}, \quad \mathbf{x_0} = \begin{bmatrix} g_0 \\ h_0 \end{bmatrix}.$$

The control variable u corresponds to the insulin infusion rate.

The cost functional (2.14) corresponds to the general quadratic cost functional (2.17), where

$$Q = 2\begin{bmatrix} 1 & 0 \\ 0 & 0 \end{bmatrix}, \quad R = 2\rho, \quad \boldsymbol{l} = \begin{bmatrix} l \\ 0 \end{bmatrix}, \quad \mathbf{e}(t) = \mathbf{x}(t) - \boldsymbol{l}.$$

We consider the infinite-horizon case ($\tau \to \infty$).

Since our goal is to minimize $\mathcal{J}$, rather than compute its exact value, the factor $\frac{1}{2}$ in front of the cost functional has no effect on the optimal control and can be omitted. To keep the structure consistent with the general LQR form, we multiplied both Q and R by 2.

The Hamiltonian function (2.18) for this system becomes

$$H = (g - l)^2 + \rho u^2 + \lambda_1(-ag - bh) + \lambda_2(-ch + dg + u) \tag{2.29}$$

where λ_1 and λ_2 are adjoint variables corresponding to state variables $g(t)$ and $h(t)$, respectively.

We now apply the general step-by-step algorithm introduced in Sect. 2.3.2 to the glucose-insulin regulation model.

Step 1

Let us turn our attention to the Riccati equation (2.27), where the unknown is the symmetric positive semidefinite matrix $P = \begin{bmatrix} q & z \\ z & j \end{bmatrix}$.

In general, the Riccati equation is a matrix differential equation that must be solved by integrating backward in time from $t = \tau$ to $t = 0$. Solving it over a finite time interval can be computationally demanding, and numerical algorithms such as the forward–backward sweep method are typically used (see Sect. 2.3.4). However, as will be shown below, in the special case of an infinite time horizon ($\tau \to \infty$) the solution may converge to a steady state.

When the matrices A, B, Q, R are constant, this steady-state behavior implies that $P(t)$ becomes time-independent and, therefore, $\frac{dP}{dt} = 0$. As a result, the Riccati equation reduces to the algebraic Riccati equation:

$$-PA - A^T P + PBR^{-1}B^T P - Q = 0.$$

For our glucose-insulin system, substituting the specific matrices gives:

$$\begin{bmatrix} q & z \\ z & j \end{bmatrix}\begin{bmatrix} -a & -b \\ d & -c \end{bmatrix} + \begin{bmatrix} -a & d \\ -b & -c \end{bmatrix}\begin{bmatrix} q & z \\ z & j \end{bmatrix} - \frac{1}{2\rho}\begin{bmatrix} q & z \\ z & j \end{bmatrix}\begin{bmatrix} 0 \\ 1 \end{bmatrix}\begin{bmatrix} 0 & 1 \end{bmatrix}\begin{bmatrix} q & z \\ z & j \end{bmatrix} + 2\begin{bmatrix} 1 & 0 \\ 0 & 0 \end{bmatrix} = \begin{bmatrix} 0 & 0 \\ 0 & 0 \end{bmatrix}.$$

Following [11] we introduce the new variables $Y_0 = \frac{q}{2\rho}$, $Y_1 = \frac{z}{2\rho}$, $Y_2 = \frac{j}{2\rho}$ to simplify the algebra and extract scalar equations from algebraic Riccati equation. Since P is a symmetric matrix, there will be only three equations:

$$2aY_0 - 2dY_1 + Y_1^2 - \frac{1}{\rho} = 0 \tag{2.30a}$$

$$bY_0 + Y_1(Y_2 + a + c) - dY_2 = 0 \tag{2.30b}$$

$$2bY_1 + 2cY_2 + Y_2^2 = 0 \tag{2.30c}$$

We first isolate Y_0 from the first two Eqs. (2.30), then equate the resulting expressions to get

$$bY_1^2 - (2db + 2a(a+c))Y_1 - 2aY_1Y_2 + 2adY_2 - \frac{b}{\rho} = 0$$

We now isolate Y_1 from the last equation (2.30) and substitute it in the expression above. As a result, we get the quadratic equation

$$y^2 + 4(ac + bd)y - \frac{4b^2}{\rho} = 0 \tag{2.31}$$

where $y = Y_2^2 + 2(a+c)Y_2$.

Solving this, we find

$$Y_2 = -(a+c) \pm \sqrt{a^2 + c^2 - 2bd \pm 2\sqrt{(ac+bd)^2 + \frac{b^2}{\rho}}}$$

To ensure that the matrix P is positive definite, we require $q > 0$, $qj - z^2 > 0$. This implies $j > 0$, so we choose the positive root for Y_2:

$$Y_2 = \sqrt{a^2 + c^2 - 2bd + 2\sqrt{(ac + bd)^2 + \frac{b^2}{\rho}}} - (a + c). \tag{2.32}$$

Once Y_2 is known, we can determine

$$Y_1 = -\frac{Y_2^2 + 2cY_2}{2b} < 0 \tag{2.33}$$

and then

$$Y_0 = \frac{\frac{1}{\rho} - Y_1^2 + 2dY_1}{2a} \tag{2.34}$$

Finally, we recover the components $q = 2\rho Y_0$, $z = 2\rho Y_1$, $j = 2\rho Y_2$ of matrix P. Thus, the matrix P is fully determined.

Step 2

Since we are considering the steady-state case, the differential equation for $\boldsymbol{\mu}$ from Eq. (2.28) becomes algebraic:

$$(A^T - PBR^{-1}B^T)\boldsymbol{\mu} = Q\boldsymbol{l}.$$

For our glucose-insulin model, substituting the specific matrices gives:

$$\left(\begin{bmatrix} -a & d \\ -b & -c \end{bmatrix} - \frac{1}{2\rho} \begin{bmatrix} q & z \\ z & j \end{bmatrix} \begin{bmatrix} 0 \\ 1 \end{bmatrix} \begin{bmatrix} 0 & 1 \end{bmatrix} \right) \boldsymbol{\mu} = 2 \begin{bmatrix} 1 & 0 \\ 0 & 0 \end{bmatrix} \begin{bmatrix} l \\ 0 \end{bmatrix}.$$

Using the definitions for Y_1, Y_2 and solving for $\boldsymbol{\mu}$, we get

$$\boldsymbol{\mu} = \frac{2l}{\xi} \begin{bmatrix} Y_2 + c \\ -b \end{bmatrix} \tag{2.35}$$

where

$$\xi = bY_1 - aY_2 - ac - bd.$$

Substituting the expressions (2.32) and (2.33) for Y_1 and Y_2, we obtain

$$\xi = -\sqrt{(ac + bd)^2 + \frac{b^2}{\rho}} < 0. \tag{2.36}$$

Step 3

Given the matrix P and the vector $\boldsymbol{\mu}$, we construct $\boldsymbol{\lambda}$ using the Eq. (2.26). Substituting this expression into the optimal control formula (2.24), we obtain $\mathbf{u} = -R^{-1}B^T(P\mathbf{x} + \boldsymbol{\mu})$, which in our case becomes

$$\mathbf{u} = -\frac{1}{2\rho}\begin{bmatrix} 0 & 1 \end{bmatrix}\left(\begin{bmatrix} q & z \\ z & j \end{bmatrix}\begin{bmatrix} g \\ h \end{bmatrix} + \frac{2l}{\xi}\begin{bmatrix} Y_2 + c \\ -b \end{bmatrix}\right)$$

Using definitions of Y_1 and Y_2, the control simplifies to the form

$$\mathbf{u} = -Y_1 g - Y_2 h + Y, \tag{2.37}$$

where $Y = \frac{bl}{\rho\xi}$. Note that $Y < 0$ since $\xi < 0$ as shown in Eq. (2.36).

Step 4

Now we can substitute the expression for the optimal control from Eq. (2.37) into (2.15) to obtain the system for $g(t)$ and $h(t)$:

$$\begin{cases} \dfrac{dg}{dt} = -ag - bh, \\ \dfrac{dh}{dt} = -(Y_1 - d)g - (c + Y_2)h + Y \\ g(0) = g_0, \quad h(0) = h_0 \end{cases} \tag{2.38}$$

To reduce this system to a single equation for $g(t)$, we eliminate $h(t)$. Differentiating the first equation and substituting $\frac{dh}{dt}$ from the second equation, we obtain the initial value problem for a second-order linear differential equation with constant coefficients:

$$\begin{cases} \dfrac{d^2g}{dt^2} + (Y_2 + a + c)\dfrac{dg}{dt} - \xi g = -bY, \\ \\ g(0) = g_0, \quad \frac{dg}{dt}(0) = -ag_0 - bh_0 \end{cases} \tag{2.39}$$

The solution of the initial value problem (2.39) can be written as

$$g(t) = (C_1 \cos(\omega t) + C_2 \sin(\omega t))\, e^{-\alpha t} + g_p, \tag{2.40}$$

where

$$g_p = \frac{l}{1 + \rho(ac + bd)^2/b^2} \tag{2.41}$$

is the particular (steady-state) solution. The parameters α and ω are given by

$$\alpha = \frac{1}{2}(Y_2 + a + c) = \frac{1}{2}\sqrt{a^2 + c^2 - 2db + 2\sqrt{(ac+bd)^2 + b^2/\rho}}, \tag{2.42a}$$

$$\omega = \frac{1}{2}\sqrt{|a^2 + c^2 - 2bd + 2\xi|} = \frac{1}{2}\sqrt{|a^2 + c^2 - 2db - 2\sqrt{(ac+bd)^2 + b^2/\rho}|} \tag{2.42b}$$

Here, α plays the role of a damping coefficient, controlling the rate at which transient deviations decay, and ω represents the frequency of transient oscillations around the steady-state solution.

The constants $C_1 = g_0 - g_p$ and $C_2 = \frac{\alpha(g_0-g_p)-ag_0-bh_0}{\omega}$ are determined from the initial conditions.

Once $g(t)$ is known, the function $h(t)$ can be recovered using the first equation in (2.38) rewritten as $h = -\frac{1}{b}\frac{dg}{dt} - \frac{a}{b}g$. Substituting expressions (2.42) into this relation yields the following result:

$$h(t) = (D_1 \cos(\omega t) + D_2 \sin(\omega t))\, e^{-\alpha t} - \frac{a}{b}g_p, \tag{2.43}$$

where

$$D_1 = -\frac{1}{b}\left((a-\alpha)C_1 + \omega C_2\right), \quad D_2 = -\frac{1}{b}\left((a-\alpha)C_2 - \omega C_1\right).$$

Step 5
With $g(t)$ and $h(t)$ known, we can use Eq. (2.37) to obtain the explicit time-dependent solution for the optimal control

$$u(t) = (E_1 \cos(\omega t) + E_2 \sin(\omega t))\, e^{-\alpha t} + E_3, \tag{2.44}$$

where

$$E_1 = -Y_1 C_1 - Y_2 D_1, \quad E_2 = -Y_1 C_2 - Y_2 D_2, \quad E_3 = \left(-Y_1 + \frac{a}{b}Y_2\right) g_p + Y$$

Our approach generalizes the case presented in Swan [11], in which the parameter d, representing the direct influence of glucose on insulin dynamics, is set to zero. This reflects the physiological context of type 1 diabetes, where the body cannot produce insulin. By allowing $d \neq 0$, our formulation accounts for glucose influencing insulin dynamics, making it applicable to both type 1 and type 2 diabetes. Setting $d = 0$ in Eqs. (2.30)–(2.42) recovers the results of [11].

To illustrate the LQR method, we consider an example that uses patient-specific parameter values. Table 2.4 lists the coefficients a, b, c, d, adopted from Yipintsoi et al. [12], who reported physiological estimates for glucose–insulin dynamics in diabetic individuals.

Table 2.4 Patient-specific parameters for glucose-insulin regulation. Values of a, b, c, d are adopted from paper [12]

Patients	a, min^{-1}	b, $\frac{mg}{\mu U \cdot min}$	c, min^{-1}	d $\frac{\mu U}{mg \cdot min}$
D4b	0.0009	0.0031	0.04115	0
D4a	0.0020	0.0014	0.0220	0
D7a	0.0027	0.0012	0.0260	0
D2	0.0016	0.0143	0.0293	0

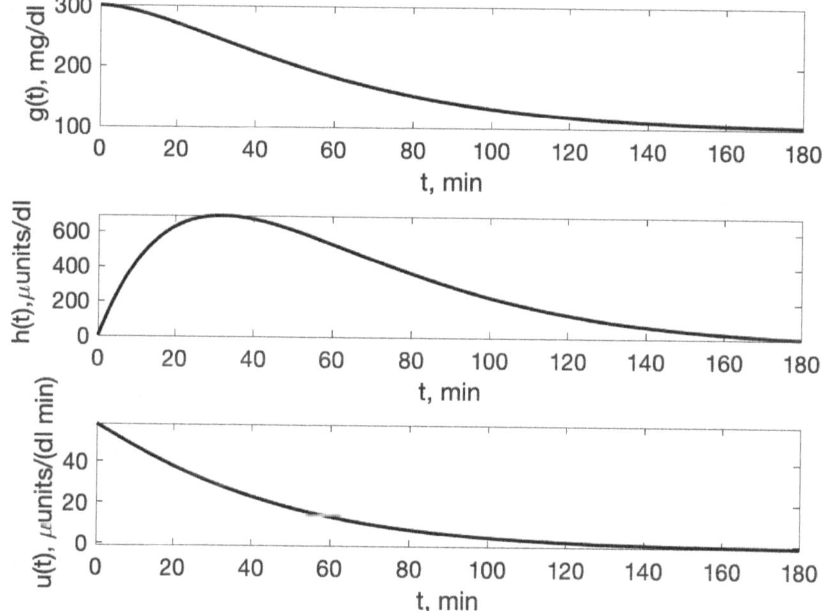

Fig. 2.14 Solution of the closed-form expressions for glucose concentration, insulin level, and infusion rate under optimal control for patient D4b. Parameter values are taken from Table 2.4 with initial glucose level $g(0) = 300$ mg/dL and target level $l = 100$ mg/dL

Figure 2.14 illustrates the implementation of Eqs. (2.40), (2.43), and (2.44) in MATLAB, showing the evolution of glucose concentration $g(t)$, insulin action $h(t)$, and insulin infusion rate $u(t)$. The simulation uses the optimal control strategy for patient D4b (Table 2.4) with initial conditions $g(0) = 300\,\text{mg/dL}$, $h(0) = 0\,\mu\text{U/mL}$, target glucose $l = 100\,\text{mg/dL}$, and weighting parameter $\rho = 10\,\text{mg}^2 \cdot \text{min}^2/\mu\text{U}^2$. Glucose $g(t)$ gradually decreases to the target over about three hours. Insulin action $h(t)$ increases as administered insulin takes effect and then returns to zero once glucose stabilizes. The control input $u(t)$ initially rises to counteract high glucose and then declines to zero as the target level is reached.

To further investigate the role of the weighting parameter ρ in the optimal control strategy, we consider the same patient D4b but vary ρ while keeping all other parameters

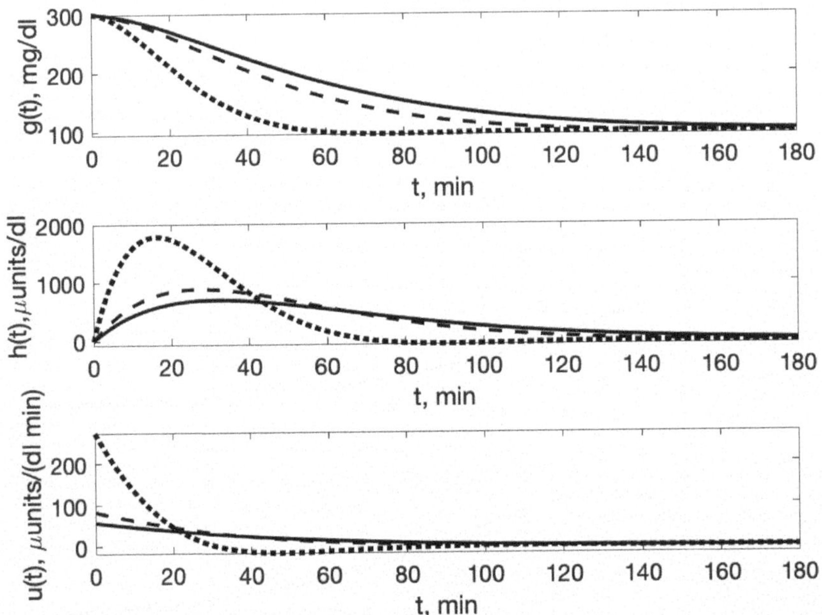

Fig. 2.15 Effect of the weighting parameter ρ on glucose concentration $g(t)$, insulin level $h(t)$ and and insulin infusion rate $u(t)$ for patient D4b. Solid curves correspond to $\rho = 10\,\frac{\text{mg}^2\cdot\text{min}^2}{\mu\text{U}^2}$, dashed curves to $\rho = 5\,\frac{\text{mg}^2\cdot\text{min}^2}{\mu\text{U}^2}$, and dotted curves to $\rho = 0.5\,\frac{\text{mg}^2\cdot\text{min}^2}{\mu\text{U}^2}$

unchanged. We compare the solutions for $\rho = 10\,\frac{\text{mg}^2\cdot\text{min}^2}{\mu\text{U}^2}$, $\rho = 5\,\frac{\text{mg}^2\cdot\text{min}^2}{\mu\text{U}^2}$, and $\rho = 0.5\,\frac{\text{mg}^2\cdot\text{min}^2}{\mu\text{U}^2}$ using the analytical formulas derived in Eqs. (2.40), (2.43), and (2.44).

As shown in Fig. 2.15, the decrease ρ reduces the penalty for insulin usage, resulting in a faster and more forceful response from the control input $u(t)$. This leads to a faster correction of the blood glucose level $g(t)$, but also produces a steeper increase in insulin activity $h(t)$, which may be less desirable physiologically. The comparison illustrates the trade-off in LQR design. Smaller ρ values produce faster glucose regulation, whereas larger values emphasize control efficiency and produce smoother, slower responses.

MATLAB codes used to generate the plots in Figs. 2.14 and 2.15 are available on GitHub [8].

The control variable $u(t)$ derived using the LQR method and given analytically in Eq. (2.44), describes a damped oscillation around a steady-state value. For certain parameter values and initial conditions, the transient behavior or even the steady-state value of the control may become negative. While such behavior is mathematically optimal within the LQR framework, negative infusion rates are not physiologically feasible. To address this, we implement the control only while it remains nonnegative. Once $u(t)$ becomes negative, it is set to zero for all subsequent times.

2.3.4 Optimal Control on a Fixed Time Interval

So far, we have focused on the case, where the time interval extends to infinity and the solution of the Riccati equation reduces to an algebraic form. This setting allowed for closed-form expressions of the optimal control and state trajectories, reflecting steady-state behavior.

We now turn to the finite-horizon case, where the control is applied over a fixed interval $0 \le t \le \tau$. In this setting, the Riccati equation becomes a matrix differential equation that must be solved backward in time. The resulting optimal control problem is formulated as a two-point boundary value problem, involving coupled differential equations for the state and costate variables.

Given the Hamiltonian in the form of (2.29), we apply the optimality condition (2.22) to obtain the control:

$$u(t) = -\frac{\lambda_2(t)}{2\rho}. \tag{2.45}$$

Substituting this into the second state equation in (2.15), and using the costate equations (2.21) along with the transversality conditions (2.23), we arrive at the following two-point boundary value problem:

$$\frac{dg}{dt} = -ag - bh, \tag{2.46a}$$

$$\frac{dh}{dt} = -ch + dg - \frac{\lambda_2}{2\rho}, \tag{2.46b}$$

$$\frac{d\lambda_1}{dt} = -2(g - l) + a\lambda_1 - d\lambda_2, \tag{2.46c}$$

$$\frac{d\lambda_2}{dt} = b\lambda_1 + c\lambda_2, \tag{2.46d}$$

with initial conditions

$$g(0) = g_0, \quad h(0) = h_0, \tag{2.46e}$$

and terminal (transversality) conditions

$$\lambda_1(\tau) = 0, \quad \lambda_2(\tau) = 0. \tag{2.46f}$$

Closed-form solutions to (2.46) are generally not available, so numerical methods are typically required. In what follows, we illustrate the solution process using the forward–backward sweep method, a standard approach for two-point boundary value problems.

In our case, the blood glucose regulation model is linear with constant coefficients. This allows us to obtain an explicit analytical solution for the finite-horizon optimal control problem, which helps us understand the system's behavior and provides a way to check numerical results.

Analytical Solution of the Finite-Horizon Optimal Control Problem

The system of Eqs. (2.46a)–(2.46d) can be rewritten in the compact matrix form

$$\frac{d\mathbf{x}}{dt} = M\mathbf{x} + \mathbf{n}, \tag{2.47}$$

where $\mathbf{x} = [g; h; \lambda_1; \lambda_2]^T$ and

$$M = \begin{bmatrix} -a & -b & 0 & 0 \\ d & -c & 0 & -\frac{1}{2\rho} \\ -2 & 0 & a & -d \\ 0 & 0 & b & c \end{bmatrix}, \quad \mathbf{n} = \begin{bmatrix} 0 \\ 0 \\ 2l \\ 0 \end{bmatrix}. \tag{2.48}$$

Given the target glucose level l, the initial states g_0 and h_0, and the final time τ, the solution vector $\mathbf{x}(t)$ can be expressed as the sum of the homogeneous and particular components:

$$\mathbf{x}(t) = \mathbf{x}_h(t) + \mathbf{x}_p(t),$$

where $\mathbf{x}_h(t)$ satisfies the homogeneous system

$$\frac{d\mathbf{x}_h}{dt} = M\mathbf{x}_h$$

and $\mathbf{x}_p(t)$ is a particular solution of the non-homogeneous system (2.47).

To construct the homogeneous solution $\mathbf{x}_h(t)$, we need the eigenvalues and eigenvectors of the coefficient matrix M. These quantities describe the system's modes and determine the structure of the general solution.

The eigenvalues r_i and eigenvectors $\mathbf{v}_i$ $(i = 1, 2, 3, 4)$ of the matrix M can be obtained analytically or computed numerically using the `eig` function in MATLAB.

To derive them analytically, we solve the characteristic equation

$$det(M - rI_4) = 0,$$

where I_4 is the identity (4×4) matrix. In this case, the characteristic equation reduces to a biquadratic form

$$r^4 - r^2(a^2 + c^2 - 2bd) + (ac + bd)^2 + \frac{b^2}{\rho} = 0$$

with four roots. These roots can be expressed as

$$
\begin{aligned}
r_1 &= +\sqrt{\frac{A_1 + \sqrt{A_1^2 - 4A_2}}{2}}, \quad r_2 = +\sqrt{\frac{A_1 - \sqrt{A_1^2 - 4A_2}}{2}}, \\
r_3 &= -\sqrt{\frac{A_1 + \sqrt{A_1^2 - 4A_2}}{2}}, \quad r_4 = -\sqrt{\frac{A_1 - \sqrt{A_1^2 - 4A_2}}{2}},
\end{aligned}
\tag{2.49}
$$

where $A_1 = a^2 + c^2 - 2bd$, $A_2 = (ac + bd)^2 + \frac{b^2}{\rho}$.

To compute the corresponding eigenvectors $\mathbf{v}_i$, we solve the homogeneous system $(M - r_i I_4)\mathbf{v}_i = \mathbf{0}$ for each $i = 1, 2, 3, 4$. As a result, we obtain the following expressions:

$$
\mathbf{v}_i = \begin{bmatrix} \frac{1}{2}\left(a - r_i + \frac{db}{c-r_i}\right) \\ \frac{1}{c+r_i}\left(\frac{d}{2}\left(a - r_i + \frac{db}{c-r_i}\right) + \frac{b}{2\rho(c-r_i)}\right) \\ 1 \\ -\frac{b}{c-r_i} \end{bmatrix}
\tag{2.50}
$$

The homogeneous solution $\mathbf{x}_h(t)$ is a linear combination of the therms $\mathbf{v}_i e^{r_i t}$ $(i = 1, 2, 3, 4)$ with four coefficients determined from the two-point boundary conditions (2.46e)–(2.46f).

Since the vector $\mathbf{n}$ is constant, we look for a particular solution $\mathbf{x}_p$ in the form of a constant vector. Substituting $\mathbf{x}_p$ into Eq. (2.47) and setting $d\mathbf{x}_p/dt = 0$ gives the algebraic system $M\mathbf{x}_p + \mathbf{n} = 0$. Solving this system yields:

$$
\mathbf{x}_p = \begin{bmatrix} \frac{lb^2}{b^2+\rho(ac+bd)^2} \\ \frac{-abl}{b^2+\rho(ac+bd)^2} \\ \frac{-2\rho cl(ac+bd)}{b^2+\rho(ac+bd)^2} \\ \frac{2\rho bl(ac+bd)}{b^2+\rho(ac+bd)^2} \end{bmatrix}.
\tag{2.51}
$$

The first two components correspond to the steady-state glucose and insulin levels derived in (2.40) and (2.43), showing the connection between the finite-horizon solution and the steady-state (infinite-horizon) behavior.

By choosing values for l and ρ along with the parameters a, b, c, d, we can obtain the eigenvalues and eigenvectors analytically using formulas (2.49) and (2.50). For example, for $l = 100\,\text{mg/dL}$, $\rho = 10\,\frac{\text{mg}^2\cdot\text{min}^2}{\mu\text{U}^2}$, and the parameters for patient D4b in Table 2.4, we obtain $r_j = \pm 0.0304 \pm i\,0.0077$, $j = 1, 2, 3, 4$. Such complex eigenvalues are typical in control systems with coupled oscillatory dynamics, such as glucose-insulin feedback interactions.

Since M has real entries, its complex eigenvalues appear in conjugate pairs. Let the eigenvalues be $r_{1,2} = \alpha \pm i\omega$ and $r_{3,4} = -\alpha \pm i\omega$, with corresponding conjugate eigenvectors $\mathbf{v}_2 = \mathbf{v}_1^*$ and $\mathbf{v}_4 = \mathbf{v}_3^*$. In this case, the general solution to system (2.46) can be expressed in real form as:

$$\mathbf{x} = e^{\alpha t}(C^{(1)}\mathfrak{R}(\mathbf{v}_1)\cos\omega t - C^{(2)}\mathfrak{I}(\mathbf{v}_1)\sin\omega t) + e^{-\alpha t}(C^{(3)}\mathfrak{R}(\mathbf{v}_3)\cos\omega t - C^{(4)}\mathfrak{I}(\mathbf{v}_3)\sin\omega t) + \mathbf{x}_p, \tag{2.52}$$

where $\mathfrak{R}$ and $\mathfrak{I}$ are the notation for the real and imaginary parts of the vectors. Constants $C^{(1)}, C^{(2)}, C^{(3)}, C^{(4)}$ need to be found from two-point boundary conditions (2.46e)–(2.46f).

Equation (2.52) illustrates how complex conjugate eigenvalues give rise to oscillatory dynamics.

Because the solution contains both growing and decaying exponentials that can differ greatly in magnitude for moderate values of τ, the resulting system for the coefficients $C^{(i)}$, $i = 1, 2, 3, 4$ becomes very sensitive to roundoff errors. To address this, we use a more stable approach based on MATLAB's complex-valued eigenvalue decomposition.

We express the homogeneous solution as a linear combination of eigenmodes:

$$\mathbf{x}_h(t) = \sum_{i=1}^{4} C_i\, \mathbf{v}_i\, e^{r_i t},$$

where $C_i \in \mathbb{C}$, and r_i and $\mathbf{v}_i$ are the eigenvalues and eigenvectors of the matrix M calculated numerically using the built-in MATLAB function `eig`. Even though this representation involves complex-valued eigenvalues and eigenvectors, the resulting solution is real-valued because all complex terms appear in conjugate pairs and their imaginary parts cancel when combined. This allows us to solve for the constants C_i robustly using standard numerical linear algebra techniques, while still preserving analytical insight into the oscillatory dynamics of the system.

In Fig. 2.16, we plot the solution

$$\mathbf{x}(t) = \mathbf{x}_p + \sum_{i=1}^{4} C_i\, \mathbf{v}_i\, e^{r_i t},$$

for system (2.47) using the two-point boundary conditions (2.46e) and (2.46f). We show the time series for $g(t)$ and $h(t)$ (the first two components of $\mathbf{x}$). The fourth component, $\lambda_2(t)$, is used to compute the control $u(t)$ via (2.45), which is also plotted as a function of time. Parameter values are taken from Table 2.4 (Patient D4a), with target glucose $l =$

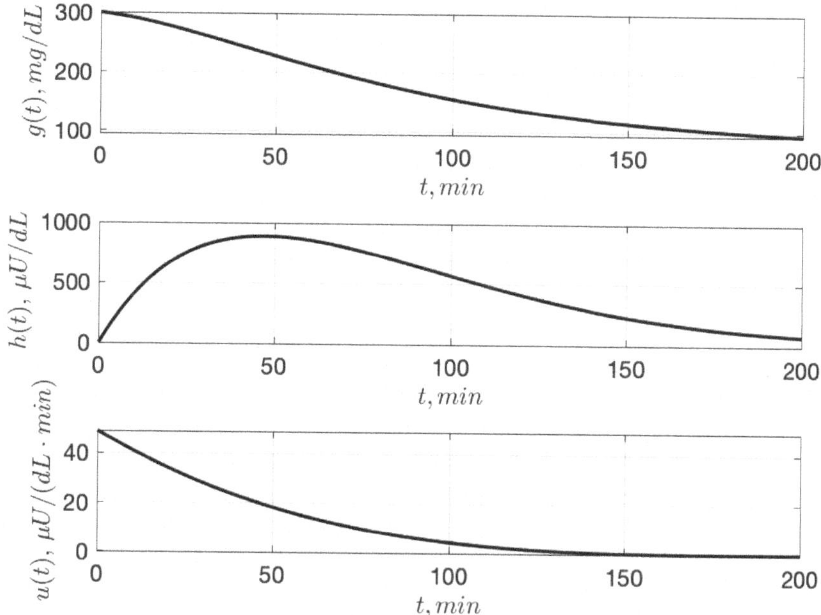

Fig. 2.16 Time series of glucose $g(t)$, insulin $h(t)$, and control $u(t)$ obtained from the analytical solution of the finite-horizon optimal control problem (2.47). Parameters are $g_0 = 300\,\text{mg/dL}$, $l = 100\,\text{mg/dL}$, $h_0 = 0\,\mu\text{U/mL}$, $\rho = 10\,\frac{\text{mg}^2\cdot\text{min}^2}{\mu\text{U}^2}$, and $\tau = 200$ min. The values of a, b, c, d correspond to Patient D4a (see Table 2.4)

$100\,\text{mg/dL}$, $\rho = 10\,\frac{\text{mg}^2\cdot\text{min}^2}{\mu\text{U}^2}$, and $\tau = 200$ minutes. MATLAB code used to generate Fig. 2.16 is available on GitHub [8].

The analytical solution of the finite-horizon problem can also be compared with the earlier infinite-horizon LQR results. For example, using the parameter set for Patient D4b with a horizon of $\tau = 180$ minutes, the finite-horizon solution closely matches the LQR trajectories.

While this analytical approach clarifies the structure of the finite-horizon problem, it does not extend to more complex models with nonlinear or time-varying dynamics. To generalize and address such cases, we now turn to a numerical approach.

Numerical Solution Using the Forward-Backward Sweep Method

To solve the finite-horizon optimal control problem numerically, we apply the forward-backward sweep method (see, e.g., [13]). This iterative technique alternates between forward integration of the state equations and backward integration of the costate equations, with the control updated at each step. Because of its stability and generality, the method is widely used for systems governed by the Pontryagin Maximum Principle.

To solve the finite-horizon optimal control problem in glucose–insulin regulation, the forward-backward sweep method proceeds as follows:

1. Make an initial guess for the control $u(t)$ over the time interval $[0, \tau]$.
2. Perform forward integration of the state equations for glucose $g(t)$ and insulin $h(t)$ using the current $u(t)$ and the initial conditions $g(0) = g_0$ and $h(0) = h_0$.
3. Perform backward integration of the adjoint equations for the costate variables $\lambda_1(t)$ and $\lambda_2(t)$ using computed values of $g(t)$ and $h(t)$ and transversality (terminal) conditions $\lambda_1(\tau) = \lambda_2(\tau) = 0$.
4. Update of the control $u(t)$ using the current value of the adjoint variable $\lambda_2(t)$ (see (2.45)).
5. Check for convergence. Compare the values of the variables between successive iterations. If the change is smaller than the chosen tolerance (for example, 10^{-4}), assume that convergence has been achieved; otherwise, return to Step 2.

Our numerical implementation is adapted from the MATLAB code presented in [13]. We have modified it to incorporate a non-zero parameter d, and it uses the physiological parameters a, b, c, d listed in Table 2.4. The code is available on GitHub [8]. Some explanatory comments from [13] are reproduced below, as they remain directly relevant to our implementation.

Step 1: Initial Guess

For Step 1, we begin with the initial guess $u = 0,\ t \in [0, \tau]$.

Steps 2 and 3: Numerical Integration of State and Adjoint Equations

For Steps 2 and 3, any numerical ODE solver can be used, for instance, MATLAB's ode45. Here, we implement the classical fourth-order Runge-Kutta (RK4) method because it provides sufficient accuracy while allowing students to engage directly with the numerical algorithm (see, e.g., the standard numerical methods textbook [14] for details).

In general, given an ordinary differential equation of the form $\frac{dx}{dt} = f(t, x(t))$, the RK4 method approximates the solution at the next time step as:

$$x(t + dt) \approx x(t) + \frac{dt}{6}(k_1 + 2k_2 + 2k_3 + k_4), \tag{2.53}$$

where

$$k_1 = f(t, x(t)) \tag{2.54a}$$

$$k_2 = f(t + 0.5 \cdot dt, x(t) + 0.5 \cdot dt \cdot k_1) \tag{2.54b}$$

$$k_3 = f(t + 0.5 \cdot dt, x(t) + 0.5 \cdot dt \cdot k_2) \tag{2.54c}$$

$$k_4 = f(t + dt, x(t) + dt \cdot k_3) \tag{2.54d}$$

The local truncation error of RK4 is $O(dt^5)$, and the global error is $O(dt^4)$. In our implementation, we choose a final time τ (measured in minutes), and divide the interval into $N = 1000$ time steps of size $dt = \frac{\tau}{N}$.

Step 2: Forward Integration of State Equations

With the control $u(t)$ and the RK4 method in place, we perform the forward sweep to integrate the state equations for glucose $g(t)$ and insulin $h(t)$ from $t = 0$ to $t = \tau$.

For our system, the state equations depend explicitly on the control $u(t)$ which changes during each iteration i, $(i = 1, 2, \ldots, N)$. When applying RK4, the evaluation of the insulin equation at intermediate points requires control in a midstep $u(t + 0.5 \cdot dt)$.

Since u is only defined at discrete time points t_i, we approximate this midpoint value using the average: $u(t + 0.5 \cdot dt) \approx 0.5(u(i) + u(i + 1))$.

This simple approximation follows the approach in [13] and is sufficient for our linear glucose–insulin system with small time steps. For rapidly changing controls or highly nonlinear systems, higher-order interpolations (quadratic, cubic) could be considered.

The forward sweep in MATLAB can then be implemented as follows:

```
% === Forward RK4 ===
for i = 1:N
    k1g = -a * g(i) - b * h(i);
    k1h = -c * h(i) + d * g(i) + u(i);

    k2g = -a * (g(i) + dt/2 * k1g) - b * (h(i) + dt/2 * k1h);
    k2h = -c * (h(i) + dt/2 * k1h) + d * (g(i) + dt/2 * k1g) +
    0.5 * (u(i) + u(i+1));

    k3g = -a * (g(i) + dt/2 * k2g) - b * (h(i) + dt/2 * k2h);
    k3h = -c * (h(i) + dt/2 * k2h) + d * (g(i) + dt/2 * k2g) +
    0.5 * (u(i) + u(i+1));

    k4g = -a * (g(i) + dt * k3g) - b * (h(i) + dt * k3h);
    k4h = -c * (h(i) + dt * k3h) + d * (g(i) + dt * k3g) + u(i+1);

    g(i+1) = g(i) + (dt / 6) * (k1g + 2*k2g + 2*k3g + k4g);
    h(i+1) = h(i) + (dt / 6) * (k1h + 2*k2h + 2*k3h + k4h);
end
```

Here g and h store the glucose and insulin values, respectively, at each time step.

Step 3: Backward Integration of the Adjoint Equations

After completion of the forward sweep for the state variables, we integrate the adjoint (costate) equations backward in time from $t = \tau$ to $t = 0$. The costate equations in our finite-horizon problem are the third and fourth equations in system (2.46), with terminal (transversality) conditions: $\lambda_1(\tau) = \lambda_2(\tau) = 0$ (see (2.46f)).

As with the forward sweep, we apply the classical RK4 method. Because the integration is backward, the effective time increment is negative ($-dt$). As with the control $u(t)$ in Step 2, we approximate the backward half-step for the variable $g(t)$ using an average $g(t - 0.5 \cdot dt) \approx 0.5(g(j) + g(j-1))$.

The backward sweep can be implemented in MATLAB as follows:

```
% === Backward RK4 ===
    for i = 1:N
        j = N + 2 - i;

        k1l1 = -2*(g(j) - 1) + a * lambda1(j) - d * lambda2(j);
        k1l2 = b * lambda1(j) + c * lambda2(j);

        k2l1 = -2*(0.5*(g(j)+g(j-1)) - 1) + a * (lambda1(j) - dt/2*k1l1)
        -d * (lambda2(j) - dt/2*k1l2);
        k2l2 = b * (lambda1(j) - dt/2*k1l1) + c * (lambda2(j) - dt/2*k1l2);

        k3l1 = -2*(0.5*(g(j)+g(j-1)) - 1) + a * (lambda1(j) - dt/2*k2l1)
        - d * (lambda2(j) - dt/2*k2l2);
        k3l2 = b * (lambda1(j) - dt/2*k2l1) + c * (lambda2(j) - dt/2*k2l2);

        k4l1 = -2*(g(j-1) - 1) + a * (lambda1(j) - dt*k3l1)
        - d * (lambda2(j) - dt*k3l2);
        k4l2 = b * (lambda1(j) - dt*k3l1) + c * (lambda2(j) - dt*k3l2);

        lambda1(j-1) = lambda1(j) - (dt/6)*(k1l1 + 2*k2l1 + 2*k3l1 + k4l1);
        lambda2(j-1) = lambda2(j) - (dt/6)*(k1l2 + 2*k2l2 + 2*k3l2 + k4l2);
    end
```

Here, the index j runs from $N + 1$ to 2 allowing us to step backward in time. By the end of the backward sweep, we have the costate trajectories $\lambda_1(t)$ and $\lambda_2(t)$ for the current iteration.

Step 4: Update the Control

After the backward sweep we have the costate $\lambda_2(t)$. The pointwise optimal control (from the optimality condition (2.45)) is $u_1(t) = -\frac{\lambda_2}{2\beta}$.

Direct substitution of this expression may cause oscillations or instability in early iterations. To improve stability and convergence, we use a weighted average, taking a combination of the new control ($u(t)$) and the previous iterate ($u_{old}(t)$):

$$u(t) = \theta u_1(t) + (1 - \theta)u_{old}(t),$$

where $0 < \theta \leq 1$ is a relaxation parameter. Following [13], we set $\theta = 0.5$, which balances convergence speed with stability. If convergence is too slow, θ can be increased; if oscillations persist, θ can be decreased.

MATLAB implementation:

```
% Update control
u1 = -lambda2 / (2 * rho);    % candidate control from optimality condition
```

```
theta = 0.5;                    % relaxation parameter
u = theta * u1 + (1 - theta) * oldu;
```

This weighted averaging smooths oscillations in successive iterations and helps the control sequence converge steadily to the optimal solution.

Step 5: Check for Convergence
After updating the control in Step 4, we need to determine whether the forward-backward sweep has converged. Convergence is assessed by comparing the values of the variables between successive iterations. Specifically, we compute the sum of absolute differences for each variable (control $u(t)$, states $g(t)$, $h(t)$ and costates $\lambda_1(t)$, $\lambda_2(t)$) and compare it to a small tolerance, chosen as $\delta = 10^{-4}$.

For example, for the control:

$$temp_1 = \delta \sum_{i=1}^{N+1} |u(i)| - \sum_{i=1}^{N+1} |u(i) - u_{old}(i)|.$$

The quantity $temp_1$ measures the relative change in the control between iterations, expressed as an l^1 norm (sum of absolute values).

We perform similar calculations for g, h, λ_1, λ_2 and then take the minimum:

$$test = min(temp_1, temp_2, temp_3, temp_4, temp_5)$$

If $test \geq 0$, the iteration has converged; otherwise, we repeat Steps 2–4.

MATLAB implementation:

```
while (test < 0)
    % Save previous iteration values
    oldu = u;
    oldg = g;
    oldh = h;
    oldlambda1 = lambda1;
    oldlambda2 = lambda2;
 % === Forward RK4 ===
 ..........
 % === Backward RK4 ===
 ..........
  % Update control
  ..........
  % Convergence check
    temp1 = delta * sum(abs(u)) - sum(abs(oldu - u));
    temp2 = delta * sum(abs(g)) - sum(abs(oldg - g));
    temp3 = delta * sum(abs(h)) - sum(abs(oldh - h));
    temp4 = delta * sum(abs(lambda1)) - sum(abs(oldlambda1 - lambda1));
    temp5 = delta * sum(abs(lambda2)) - sum(abs(oldlambda2 - lambda2));
```

```
    test = min([temp1, temp2, temp3, temp4, temp5]);
end
```

This check ensures that all variables have stabilized before terminating the forward-backward sweep, yielding a smooth and reliable approximation of the optimal control and state trajectories.

Note: We use the l^1 norm for simplicity. For stiff or highly nonlinear systems, the l^2 norm (Euclidean norm) may be preferable, as it provides a smoother measure of relative error.

Upon convergence, the program prints a summary of the final values of $g(t)$, $h(t)$ and $u(t)$ and plots them as functions of time over the interval $[0, \tau]$.

We now proceed to a numerical example to illustrate the implementation of the forward-backward sweep method. We adopt the physiological parameters a, b, c, d for patient D7a from Table 2.4. For the initial condition, we set the glucose level to $g(0) = g_0 = 300\,\text{mg/dL}$ and specify the desired target level as $l = 100\,\text{mg/dL}$. The weight parameter in the cost functional is chosen as $\rho = 10\frac{\text{mg}^2 min^2}{\mu\text{U}^2}$. We perform the simulation on the time interval $[0, \tau]$, with $\tau = 210$ minutes. The computed results for glucose $g(t)$, insulin $h(t)$, and control $u(t)$ are shown in Fig. 2.17.

In this case, the optimal control $u(t)$ is monotonic and steadily decreases to zero. This behavior reflects a decreasing need for external insulin input as the glucose level

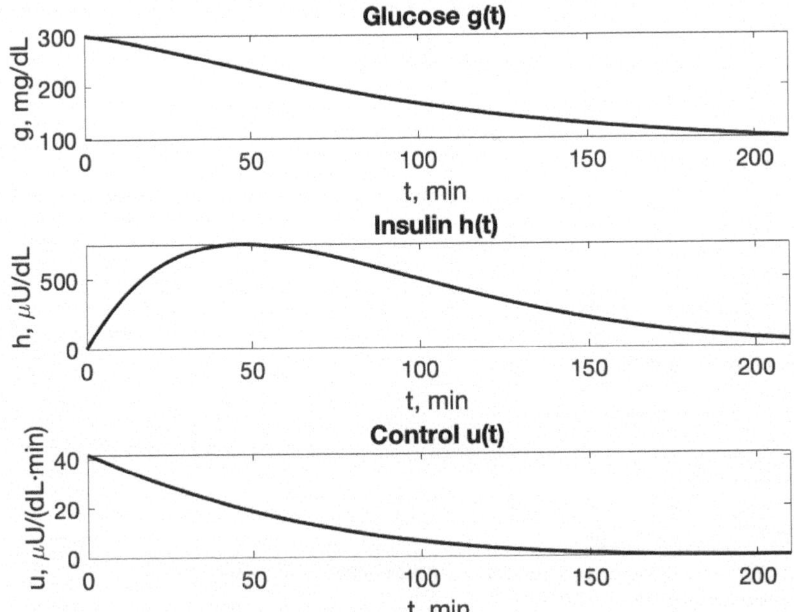

Fig. 2.17 Time evolution of glucose $g(t)$, insulin $h(t)$, and control $u(t)$ for the numerical example with $\tau = 210\,min$, $g(0) = 300\,\text{mg/dL}$, $l = 100\,\text{mg/dL}$, and $\rho = 10\frac{\text{mg}^2 min^2}{\mu\text{U}^2}$. Parameter values a, b, c, d for patient D7a are taken from Table 2.4

approaches the desired target. The insulin $h(t)$ initially increases in response to the control and then declines as $u(t)$ declines. As a result, the glucose concentration $g(t)$ gradually decreases and approaches the target level by the end of the simulation.

This example demonstrates how the forward-backward sweep method produces a smooth, convergent control $u(t)$ that drives insulin and glucose levels toward the target. It provides a clear illustration of how the numerical algorithm translates the optimality conditions into practical system behavior.

2.4 Nonlinear Modeling and Bifurcation in Diabetes Onset

This section shifts the focus from managing diabetes to modeling its onset using nonlinear differential equations. The model is adapted from the work of Topp et al. [15] and is commonly referred to as the βIG model, where β represents the mass of pancreatic β-cells, G denotes the blood glucose concentration, and I denotes the blood insulin concentration. In [16], the authors adapted the model for instructional use, emphasizing its bifurcation behavior and sensitivity to parameter changes. Bifurcation analysis is especially important in this context because it identifies parameters where small changes can trigger qualitative shifts in system dynamics, for instance, a transition from stable glucose regulation to chronic hyperglycemia. Sensitivity analysis complements this by quantifying which parameters exert the strongest influence on these transitions, thereby clarifying the mechanisms underlying diabetes onset and pointing to potential targets for medical intervention. A more detailed discussion of sensitivity analysis is provided in Sect. 2.4.2.

To maintain consistency with the notation used in the previous sections, we denote the insulin concentration by H throughout this section.

The model is based on the following assumptions.

- The change in β-cell mass $\frac{d\beta}{dt}$ is assumed to be proportional to β, with the growth rate modeled as a quadratic function of glucose concentration $-d_0 + m_1 G - m_2 G^2$ (see the first equation in (2.55)). Here d_0 is the death rate of the mass of the β cells at zero glucose; m_1 and m_2 are the constants representing the replication and death of the mass of the β cells in response to the change in the glucose level.
- All β-cells secrete insulin at the same maximal rate σ, and the secretion rate increases with glucose concentration following a sigmoidal (Hill-type) response $\frac{G^2}{\gamma+G^2}$ ranging from 0 to 1 and reaching the half of its maximum at $G = \gamma^{1/2}$. Due to metabolic process, insulin is removed from the system with the constant rate of removal κ.
- Glucose is produced at a constant rate (modeled by ν). Glucose removal includes an insulin-independent process proportional to G with coefficient η and an insulin-dependent term proportional to HG, where coefficient of proportionality ζ reflects the body's sensitivity to insulin. Higher ζ means stronger glucose uptake for a given insulin level.

- The model assumes a single compartment for blood glucose and insulin dynamics. It ignores spatial distribution, delays, or organ-specific dynamics.
- The model does not include external glucose intake or insulin administration.

As a result, the model can be written as a system of ordinarily differential equations

$$\begin{cases} \dfrac{d\beta}{dt} &= (-d_0 + m_1 G - m_2 G^2)\beta, \\ \dfrac{dH}{dt} &= \frac{\sigma G^2}{\gamma + G^2}\beta - \kappa H, \\ \dfrac{dG}{dt} &= \nu - \eta G - \zeta H G. \end{cases} \tag{2.55}$$

for three dependent variables β, H, G, independent variable t, and nine parameters $d_0, m_1, m_2, \sigma, \gamma, \kappa, \nu, \eta, \zeta$ (see Table 2.5).

2.4.1 Equilibrium Solutions and Stability

To understand the long-term behavior of the system (2.55), we first identify its equilibrium points and then analyze their stability.

Equilibrium points correspond to states where all variables remain constant over time. To find them, we set the derivatives in system (2.55) equal to zero, getting the following system of algebraic equations:

$$F_1 = (-d_0 + m_1 G - m_2 G^2)\beta = 0, \tag{2.56a}$$

$$F_2 = \frac{\sigma G^2}{\gamma + G^2}\beta - \kappa H = 0, \tag{2.56b}$$

$$F_3 = \nu - \eta G - \zeta H G = 0 \tag{2.56c}$$

Table 2.5 Parameter values adapted from [15]

Parameter	Value	Units
d_0	0.06	d^{-1}
m_1	$0.84 \cdot 10^{-3}$	$(mg \cdot d)^{-1}dL$
m_2	$0.24 \cdot 10^{-5}$	$mg^{-2}dL^2d^{-1}$
σ	43.20	$\mu U(mg \cdot mL \cdot d)^{-1}$
γ	20,000.00	mg^2dL^{-2}
κ	432.00	d^{-1}
ν	864.00	$mg(dL \cdot d)^{-1}$
η	1.44	d^{-1}
ζ	0.72	$mL(\mu U \cdot d)^{-1}$

Assuming $\beta = 0$ in the first two equations of (2.56) implies $H = 0$. Substituting these values into the third equation yields a blood glucose concentration of $\frac{\nu}{\eta}$. Thus, one equilibrium solution is $(\beta_{e0}, H_{e0}, G_{e0}) = \left(0, 0, \frac{\nu}{\eta}\right)$.

Let $\beta \neq 0$ and suppose $m_1^2 > 4m_2d_0$. It follows from (2.56a) that in this case, the quadratic equation $m_2G^2 - m_1G + d_0 = 0$ has two distinct real roots, given by

$$G_{e1,2} = \frac{m_1 \pm \sqrt{m_1^2 - 4m_2d_0}}{2m_2}. \tag{2.57}$$

The corresponding equilibrium values of blood insulin concentration and β-cell mass are:

$$H_{e1,2} = \frac{\nu - \eta G_{e1,2}}{\zeta G_{e1,2}}, \tag{2.58}$$

$$\beta_{e1,2} = \frac{\kappa(\nu - \eta G_{e1,2})(\gamma + G_{e1,2}^2)}{\zeta \sigma G_{e1,2}^3}. \tag{2.59}$$

Therefore, this case yields two distinct equilibrium solutions: $(\beta_{e1}, H_{e1}, G_{e1})$ and $(\beta_{e2}, H_{e2}, G_{e2})$, corresponding to the negative and positive roots of (2.57), respectively.

If $\beta \neq 0$ and $m_1^2 = 4m_2d_0$, then the quadratic equation has a repeated real root, and the equilibrium solution becomes $(\beta_{e3}, H_{e3}, G_{e3})$, where:

$$G_{e3} = \frac{m_1}{2m_2}, \tag{2.60}$$

$$H_{e3} = \frac{2m_2\nu - \eta m_1}{\zeta m_1}, \tag{2.61}$$

$$\beta_{e3} = \frac{\kappa(2m_2\nu - \eta m_1)(4m_2^2\gamma + m_1^2)}{\zeta \sigma m_1^3}. \tag{2.62}$$

Note that this equilibrium reduces to $(\beta_{e0}, H_{e0}, G_{e0})$ when $m_1 = 2m_2 \cdot \frac{\nu}{\eta}$.

If $m_1^2 < 4m_2d_0$, the quadratic equation yields complex roots for G, which are not physiologically meaningful, as glucose concentration must be positive and real-valued.

Our analysis shows that the number of equilibrium solutions depend on the parameter values, particularly the relationship between m_1, m_2, and d_0.

Local stability of the equilibrium solutions is assessed by linearizing system (2.55) at each equilibrium point. To carry this out, we compute the Jacobian matrix by taking partial derivatives of the right-hand sides with respect to the system's variables:

$$J(\beta, I, G) = \begin{bmatrix} \frac{\partial F_1}{\partial \beta} & \frac{\partial F_1}{\partial H} & \frac{\partial F_1}{\partial G} \\ \frac{\partial F_2}{\partial \beta} & \frac{\partial F_2}{\partial H} & \frac{\partial F_2}{\partial G} \\ \frac{\partial F_3}{\partial \beta} & \frac{\partial F_3}{\partial H} & \frac{\partial F_3}{\partial G} \end{bmatrix} = \begin{bmatrix} -d_0 + m_1G - m_2G^2 & 0 & (m_1 - 2m_2G)\beta \\ \frac{\sigma G^2}{\gamma + G^2} & -\kappa & \frac{2G\gamma\beta\sigma}{(\gamma + G^2)^2} \\ 0 & -\zeta G & -(\eta + \zeta H) \end{bmatrix}. \tag{2.63}$$

The eigenvalues r of the Jacobian matrix $J(\beta_e, H_e, G_e)$ evaluated at a given equilibrium point (β_e, H_e, G_e) determine the local behavior of the system near that equilibrium.

These eigenvalues are the roots of the cubic characteristic equation

$$\det\left(J(\beta_e, H_e, G_e) - rI_3\right) = 0,$$

where I_3 is the 3×3 identity matrix.

By the Fundamental Theorem of Algebra, the cubic characteristic equation has three roots, which may be real or complex, and may include repeated values. If all eigenvalues have negative real parts, the equilibrium is locally asymptotically stable. If at least one eigenvalue has a positive real part, the equilibrium is unstable. If some eigenvalues have negative real parts and others are zero, the equilibrium is semistable.

We will apply this criterion to assess the stability of each equilibrium in our system.

Let us begin with the **equilibrium** $(\beta_{e0}, H_{e0}, G_{e0}) = \left(0, 0, \frac{\nu}{\eta}\right)$.

At this point, the Jacobian matrix (2.63) becomes

$$J\left(0, 0, \frac{\nu}{\eta}\right) = \begin{bmatrix} -d_0 + m_1\frac{\nu}{\eta} - m_2\frac{\nu^2}{\eta^2} & 0 & 0 \\ \frac{\sigma\frac{\nu^2}{\eta^2}}{\gamma + \frac{\nu^2}{\eta^2}} & -\kappa & 0 \\ 0 & -\zeta\frac{\nu}{\eta} & -\eta \end{bmatrix}.$$

Since this matrix is lower triangular, the characteristic polynomial is the product of terms on the diagonal of $J - rI_3$. The resulting characteristic equation is:

$$(\eta + r)(\kappa + r)\left(-d_0 + \frac{\nu(m_1\eta - m_2\nu)}{\eta^2} - r\right) = 0,$$

which yields the eigenvalues:

$$r_1 = -\eta < 0, \tag{2.64}$$

$$r_2 = -\kappa < 0, \tag{2.65}$$

$$r_3 = -d_0 + \frac{\nu(m_1\eta - m_2\nu)}{\eta^2}. \tag{2.66}$$

From expressions (2.64)–(2.66), we see that the stability of equilibrium $(\beta_{e0}, H_{e0}, G_{e0})$ depends on the sign of the eigenvalue r_3. If $r_3 < 0$, the equilibrium is locally asymptotically stable (a stable node). If $r_3 > 0$, it is unstable (a saddle point). If $r_3 = 0$, linearization is inconclusive, and the equilibrium is said to be non-hyperbolic. In this case, stability must be assessed using nonlinear analysis [20,21] or verified numerically through simulation.

The sign of r_3 depends on the parameter values used in Eq. (2.66). Using the parameter values from Table 2.5, the equilibrium $(\beta_{e0}, H_{e0}, G_{e0}) = (0, 0, \frac{\nu}{\eta})$ has the numerical value

$(0, 0, 600)$. Evaluating the Jacobian at this point, we find that all eigenvalues are real and negative:

$$r_1 = -1.44, \quad r_2 = -432.00, \quad r_3 = -0.42.$$

This indicates that the equilibrium $(0, 0, 600)$ is a stable sink. We refer to this as a "pathological equilibrium", since both the β-cell mass and blood insulin concentration are zero, while the blood glucose level is high.

At equilibrium points $(\beta_{ei}, H_{ei}, G_{ei}), i = 1, 2, 3$, the Jacobian (2.63) takes the form

$$J(\beta_{ei}, H_{ei}, G_{ei}) = \begin{bmatrix} 0 & 0 & (m_1 - 2m_2 G_{ei})\beta_{ei} \\ \frac{\sigma G_{ei}^2}{\gamma + G_{ei}^2} & -\kappa & \frac{2G_{ei}\gamma\beta_{ei}\sigma}{(\gamma + G_{ei}^2)^2} \\ 0 & -\zeta G_{ei} & -(\eta + \zeta H_{ei}) \end{bmatrix}, i = 1, 2, 3.$$

The characteristic equation $det(J(\beta_{ei}, H_{ei}, G_{ei}) - rI_3) = 0$ written for any of these three equilibrium points is a cubic equation.

$$r^3 + a_2 r^2 + a_1 r + a_0 = 0 \tag{2.67}$$

with coefficients

$$a_2 = \eta + \zeta H_{ei} + \kappa > 0 \tag{2.68}$$

$$a_1 = \kappa(\eta + \zeta H_{ei}) + \frac{2\gamma\sigma\zeta\beta_{ei} G_{ei}^2}{(\gamma + G_{ei}^2)^2} \tag{2.69}$$

$$a_0 = (m_1 - 2m_2 G_{ei}) \frac{\sigma\zeta\beta_{ei} G_{ei}^3}{\gamma + G_{ei}^2} \tag{2.70}$$

Let us consider **the equilibrium solutions** $(\beta_{e1,2}, H_{e1,2}, G_{e1,2})$, defined by Eqs. (2.57)–(2.59). To assess their local stability, we analyze the roots of the characteristic equation (2.67).

To gain insight into the nature of the real roots, we apply Descartes' rule of signs, which states:

If the polynomial function $\mathcal{P}(r) = a_n r^n + a_{n-1} r^{n-1} + \ldots + a_2 r^2 + a_1 r + a_0$ has real coefficients and $a_0 \neq 0$, then:

(a) The number of positive real zeros in the polynomial is equal to the number of variations in sign of $\mathcal{P}(r)$ or less than that number by an even integer.
(b) The number of negative real zeros of the polynomial is either equal to the number of variations of sign of $\mathcal{P}(-r)$ or less than that number by an even integer.

Given that $a_2 > 0$ (from (2.68)), the sign patterns of a_1 and a_0 lead to the following possibilities:

- $a_1 > 0, a_0 > 0 \Rightarrow$ no positive zeros, three or one negative zeros;
- $a_1 > 0, a_0 < 0 \Rightarrow$ one positive zero, two or zero negative zeros;
- $a_1 < 0, a_0 > 0 \Rightarrow$ two or zero positive zeros, one negative zero;
- $a_1 < 0, a_0 < 0 \Rightarrow$ one positive zero; two or zero negative zeros.

Thus, depending on parameter values, the system may exhibit either a stable sink or an unstable saddle.

If the characteristic equation (2.67) has complex roots, they must occur as conjugate pairs $r_{1,2} = a \pm i\omega$ with the third root $\lambda_3 \in \mathbb{R}$. In that case, we can factor the characteristic equation (2.67) as

$$r^3 + a_2 r^2 + a_1 r + a_0 = (r - (a + i\omega))(r - (a - i\omega))(r - r_3) = 0.$$

Expanding and matching coefficients yields:

$$a_2 = -r_3 - 2a \tag{2.71}$$

$$a_1 = \omega^2 + a^2 + 2ar_3 \tag{2.72}$$

$$a_0 = -(\omega^2 + a^2)r_3 \tag{2.73}$$

The coefficients a_2, a_1, a_0 in Eqs. (2.71)–(2.73) are the same as those defined earlier in Eqs. (2.68)–(2.70), but now expressed in terms of the roots of the characteristic polynomial. As seen in Eqs. (2.68)–(2.70), these coefficients depend on the system's parameters. Consequently, the nature of the roots and the stability of the equilibrium varies with parameter values. When a parameter reaches a critical (bifurcation) value, the system's qualitative behavior can change abruptly. If all eigenvalues have negative real parts, the equilibrium is a stable (sink or a spiral sink). If at least one eigenvalue has a positive real part, the equilibrium becomes unstable (a saddle or a spiral source). If any eigenvalue has zero real part, the equilibrium is non-hyperbolic. In particular, if a pair of complex conjugate eigenvalues crosses the imaginary axis, a Hopf bifurcation may occur, giving rise to a limit cycle ([20, 21]).

In the scenario where $\beta \neq 0$ and $m_1 = 2\sqrt{m_2 d_0}$, with the condition $m_1 \neq 2m_2\frac{\nu}{\eta}$, **the equilibrium point is** $(\beta_{e3}, H_{e3}, G_{e3})$, where equilibrium values $G_{e3}, H_{e3}, \beta_{e3}$ are defined by Eqs. (2.60)–(2.62).

Substituting equations (2.60)–(2.62) into Eqs. (2.68)–(2.70), we find that $a_0 = 0$, and characteristic equation (2.67) simplifies to

$$r(r^2 + a_2 r + a_1) = 0 \tag{2.74}$$

This equation has at least one root equal to zero ($r_3 = 0$), indicating that $(\beta_{e3}, H_{e3}, G_{e3})$ is a non-hyperbolic equilibrium point. Since $a_2 > 0$ by Eq. (2.67), the remaining two roots r_1 and r_2 of the characteristic equation (2.74) will either both have negative real parts if $a_1 > 0$. or one may be positive and the other negative if $a_1 < 0$.

Later we demonstrate that as a specific parameter varies, the two equilibria $(\beta_{e1}, H_{e1}, G_{e1})$ and $(\beta_{e2}, H_{e2}, G_{e2})$ coalesce into $(\beta_{e3}, H_{e3}, G_{e3})$. This type of behavior is typical of a saddle-node bifurcation.

Using the parameter values from Table 2.5, Eqs. (2.57)–(2.59) yield two equilibria, $(\beta_{e1}, H_{e1}, G_{e1}) = (300, 10, 100)$ and $(\beta_{e2}, H_{e2}, G_{e2}) = (37, 2.80, 250)$, while Eqs. (2.60)–(2.62) give the third equilibrium, $(\beta_{e3}, H_{e3}, G_{e3}) = (80.29, 4.86, 175.00)$.

To assess the stability of these equilibria, we compute the eigenvalues of the Jacobian at each point using the same parameter values from Table 2.5.

At $(\beta_{e1}, H_{e1}, G_{e1}) = (300, 10, 100)$, all eigenvalues of $J(300, 10, 100)$ are negative: $r_1 \approx -421.97, r_2 \approx -0.01$, and $r_3 \approx -18.66$, indicating that this equilibrium is a stable sink, referred to as the physiological equilibrium.

At the equilibrium $(\beta_{e2}, H_{e2}, G_{e2}) = (37, 2.80, 250)$, the eigenvalues of the Jacobian $J(37, 2.80, 250)$ are $r_1 \approx -431.01, r_2 \approx 0.04$ and $r_3 \approx -4.49$. Because one eigenvalue is positive, this equilibrium is an unstable saddle.

Finally, the third equilibrium $(\beta_{e3}, H_{e3}, G_{e3}) = (80.29, 4.86, 175.00)$ is non-hyperbolic. In Sect. 2.4.3 we show that small perturbations away from this point lead the system to transition toward the pathological equilibrium state.

2.4.2 Parameter Sensitivity Analysis

Varying the system's parameters reveals how changes influence its dynamic behavior. As noted earlier, parameter values affect the stability of equilibrium solutions. In this section, we analyze the sensitivity of the physiological equilibrium to parameter variations, offering insight into how such changes may contribute to the onset of diabetes.

As a starting point, we use the parameter set presented in Table 2.5. With these values, the system exhibits three equilibrium points: $(\beta_{e0}, H_{e0}, G_{e0}) = (0, 0, 600)$ (pathological equilibrium; stable), $(\beta_{e1}, H_{e1}, G_{e1}) = (300, 10, 100)$ (physiological equilibrium; stable), and $(\beta_{e2}, H_{e2}, G_{e2}) = (37, 2.80, 250)$ (unstable saddle).

Our goal is to assess how small changes in parameter values influence the physiological equilibrium. To this end, we apply the direct differential method (e.g., [17, 18], a local technique for parameter sensitivity analysis.

We begin by introducing the key idea behind this method. First, we rewrite system (2.55) in vector form:

$$\frac{d\boldsymbol{x}}{dt} = \boldsymbol{F}(\boldsymbol{x}, \boldsymbol{p}), \quad \boldsymbol{x} = \begin{bmatrix} \beta \\ I \\ G \end{bmatrix}, \quad \boldsymbol{F} = \begin{bmatrix} (-d_0 + m_1 G - m_2 G^2)\beta \\ \frac{\sigma G^2}{\gamma + G^2}\beta - \kappa H, \\ \nu - \eta G - \zeta H G. \end{bmatrix}, \tag{2.75}$$

where $\boldsymbol{p} = [d_0,\ m_1,\ m_2,\ \sigma,\ \gamma,\ \kappa,\ \nu,\ \eta,\ \zeta]^T$ is a nine-dimensional parameter vector.

We define the sensitivity vector $\boldsymbol{s}_j$ for parameter p_j as $\boldsymbol{s}_j = [s_{1,j}; s_{2,j}; s_{3,j}] = \left[\frac{\partial \beta}{\partial p_j}; \frac{\partial H}{\partial p_j}; \frac{\partial G}{\partial p_j}\right]$, where $j = 1, 2, \ldots, 9$. Each entry of $\boldsymbol{s}_j$ represents how the state variables β, H, and G respond to small changes in the parameter p_j. For example, $\frac{\partial G}{\partial m_1}$ characterizes how changes in the parameter m_1 affect blood glucose concentration G. A large value indicates high sensitivity, meaning that even small changes in m_1 can significantly impact glucose levels.

To capture the sensitivities with respect to all parameters, we define the sensitivity matrix $\boldsymbol{S}$, whose columns are $\boldsymbol{s}_j$:

$$\boldsymbol{S} = \begin{bmatrix} \frac{\partial \beta}{\partial d_0} & \frac{\partial \beta}{\partial m_1} & \frac{\partial \beta}{\partial m_2} & \frac{\partial \beta}{\partial \sigma} & \frac{\partial \beta}{\partial \gamma} & \frac{\partial \beta}{\partial \kappa} & \frac{\partial \beta}{\partial \nu} & \frac{\partial \beta}{\partial \eta} & \frac{\partial \beta}{\partial \zeta} \\ \frac{\partial H}{\partial d_0} & \frac{\partial H}{\partial m_1} & \frac{\partial H}{\partial m_2} & \frac{\partial H}{\partial \sigma} & \frac{\partial H}{\partial \gamma} & \frac{\partial H}{\partial \kappa} & \frac{\partial H}{\partial \nu} & \frac{\partial H}{\partial \eta} & \frac{\partial H}{\partial \zeta} \\ \frac{\partial G}{\partial d_0} & \frac{\partial G}{\partial m_1} & \frac{\partial G}{\partial m_2} & \frac{\partial G}{\partial \sigma} & \frac{\partial G}{\partial \gamma} & \frac{\partial G}{\partial \kappa} & \frac{\partial G}{\partial \nu} & \frac{\partial G}{\partial \eta} & \frac{\partial G}{\partial \zeta} \end{bmatrix} \tag{2.76}$$

Ideally, if exact solutions for β, H, and G were available, each component of the matrix $\boldsymbol{S}$ could be directly calculated. Since explicit solutions to the model are not available, we instead compute how these sensitivities evolve over time using a system of ODEs for $\boldsymbol{S}$:

$$\frac{d\boldsymbol{S}}{dt} = \boldsymbol{J}\boldsymbol{S} + \boldsymbol{f}, \tag{2.77}$$

where $\boldsymbol{J}$ represents the Jacobian defined by formula (2.63) and $\boldsymbol{f}$ is the matrix consisting of the derivatives of the right-hand sides of the equations of the system (2.75) with respect to the parameters:

$$\boldsymbol{f} = \begin{bmatrix} -\beta & G\beta & -G^2\beta & 0 & 0 & 0 & 0 & 0 & 0 \\ 0 & 0 & 0 & \frac{\beta G^2}{(\gamma+G^2)} & \frac{-\beta\sigma G^2}{(\gamma+G^2)^2} & -H & 0 & 0 & 0 \\ 0 & 0 & 0 & 0 & 0 & 0 & 1 & -G & -HG \end{bmatrix} \tag{2.78}$$

The system of differential equations (2.77) defines the change of sensitivity measures over time.

Now we formulate the initial value problem that includes the system of differential equations (2.75) and (2.77), together with their initial conditions:

$$\begin{cases} \frac{d\boldsymbol{x}}{dt} &= \boldsymbol{F}(\boldsymbol{x}; \boldsymbol{p}), \\ \frac{d\boldsymbol{S}}{dt} &= \boldsymbol{J}\boldsymbol{S} + \boldsymbol{f}, \\ \boldsymbol{x}(0) = & \boldsymbol{x}_0, \\ \boldsymbol{S}(0) = & \boldsymbol{S}_0 \end{cases} \tag{2.79}$$

According to [17], if no parameters appear in the initial conditions of the state equations, then $\boldsymbol{S}_0 = \boldsymbol{0}$.

Since the parameters and variables in our model may have different units and magnitudes, we introduce a dimensionless quantity called the relative sensitivity, denoted by S_{ij} and defined as $S_{ij} = \frac{\partial x_i}{\partial p_j}\frac{p_j}{x_i}$. This quantity measures the relative change in x_i with respect to a relative change in p_i. For example, $S_{12} = \frac{\partial \beta}{\partial m_1}\frac{m_1}{\beta}$, $S_{22} = \frac{\partial H}{\partial m_1}\frac{m_1}{I}$, $S_{32} = \frac{\partial G}{\partial m_1}\frac{m_1}{G}$, and so on.

Because these sensitivities vary over time, we define a sensitivity index $||S_{ij}||_2$, representing the overall magnitude of the relative sensitivity:

$$||S_{ij}||_2 = \sqrt{\sum_{k=1}^{N}\left(\frac{\partial x_i}{\partial p_j}(t_k)\frac{p_j}{x_i(t_k)}\right)^2} \tag{2.80}$$

This index allows us to compare the overall influence of each parameter on system variables over a given time interval. For example, $||S_{12}||_2 = \left\|\frac{\partial \beta}{\partial m_1}\frac{m_1}{\beta}\right\|_2 = \sqrt{\sum_{k=1}^{N}\left(\frac{\partial \beta}{\partial m_1}(t_k)\frac{m_1}{\beta(t_k)}\right)^2}$, $||S_{22}||_2 = \left\|\frac{\partial H}{\partial m_1}\frac{m_1}{H}\right\|_2 = \sqrt{\sum_{k=1}^{N}\left(\frac{\partial H}{\partial m_1}(t_k)\frac{m_1}{H(t_k)}\right)^2}$, $||S_{32}||_2 = \left\|\frac{\partial G}{\partial m_1}\frac{m_1}{G}\right\|_2 = \sqrt{\sum_{k=1}^{N}\left(\frac{\partial G}{\partial m_1}(t_k)\frac{m_1}{G(t_k)}\right)^2}$, and so on.

We can now explore how changes in specific parameters affect the behavior of β, H, and G over time. As an illustration, we examine how variations in the parameter m_1 influence the system dynamics:

$$\begin{cases} \frac{d\beta}{dt} = (-d_0 + m_1 G - m_2 G^2)\beta \\ \frac{dH}{dt} = \frac{\sigma G^2}{\gamma + G^2}\beta - \kappa H \\ \frac{dG}{dt} = \nu - \eta G - \zeta H G \\ \frac{ds_{12}}{dt} = (-d_0 + m_1 G - m_2 G^2)s_{12} + (m_1 - 2m_2 G)\beta s_{32} + G\beta \\ \frac{ds_{22}}{dt} = \frac{\sigma G^2}{\gamma + G^2}s_{12} - \kappa s_{22} + \frac{2G\gamma\beta\sigma}{(\gamma + G^2)^2}s_{32} \\ \frac{ds_{32}}{dt} = -\zeta G s_{22} - (\eta + \zeta H)s_{32} \\ \beta(0) = \beta_0, \quad H(0) = H_0, \quad G(0) = G_0 \\ s_{12}(0) = s_{22}(0) = s_{32}(0) = 0 \end{cases} \tag{2.81}$$

The first three differential equations in (2.81) reproduce the original system (2.55), while the last three are derived from the system (2.77), using the Jacobian and the second column of the matrix $\boldsymbol{f}$, since m_1 is the second parameter in our nine-dimensional parameter vector.

Similar initial value problems can be formulated to examine the sensitivities with respect to any other parameter.

We now apply the theoretical framework developed above to investigate how variations in model parameters influence the system's behavior. Figure 2.18 shows the time-dependent relative sensitivities of the state variables β, H, and G with respect to each of the nine model parameters. The initial conditions were set to $\beta(0) = 100$ mg, $H(0) = 5$ μU mL^{-1}, and $G(0) = 80$ mg dL^{-1}. Analysis of the sensitivity measures reveals that the parameters m_1, d_0, and m_2 exhibit the strongest influence on system behavior. Small perturbations in any of these three parameters significantly affect the trajectories of β, H, and G. In particular, increasing m_1 leads to a reduction in blood glucose concentration (G), driven by enhanced levels of β and insulin (H). Likewise, decreasing d_0 or m_2 produces a similar glucose-lowering effect. These observations emphasize the key role that parameters governing β-cell dynamics play in regulating glucose-insulin balance.

To systematically rank parameter influence, we computed sensitivity indices for all nine parameters (Table 2.6) and evaluated their values at $t = 300$ days, a time by which the system's transient dynamics have largely stabilized. The resulting rankings, illustrated in Fig. 2.19, remain consistent for other late time points (e.g. $t = 200$ days) indicating robustness in parameter influence over the long term. As expected, the parameters m_1, d_0, and m_2 exhibit the strongest effects, followed by ζ (insulin sensitivity) and ν (glucose production rate).

Glucose production and removal are governed by ν and η, respectively. Analysis of Fig. 2.18 indicates that variations in ν have a more pronounced effect. The sensitivity analysis shows that increasing ν or decreasing η raises the levels of β and H needed to maintain a given glucose level.

Figure 2.18 also suggests that increasing k or α, or decreasing σ, shifts the equilibrium toward higher β values, indicating additional ways in which the system compensates for parameter changes.

As shown in Fig. 2.18, the variable ζ has a negative sensitivity with respect to both β and H, indicating that a decrease in insulin sensitivity requires an increase in insulin secretion and β-cell mass to regulate glucose. This is consistent with experimental findings reported by Topp et al. [15].

It is worth noting that the most influential parameters (m_1, m_2, and d_0) correspond to internal regulatory mechanisms and are unlikely to be significantly modified through lifestyle changes. This suggests that effective intervention may require medical treatment. In contrast, insulin sensitivity (ζ) can often be influenced externally. For example, studies such as Hernandez et al. [19] show that physical exercise can increase ζ by up to 36%, while age and obesity may reduce it by as much as 60%.

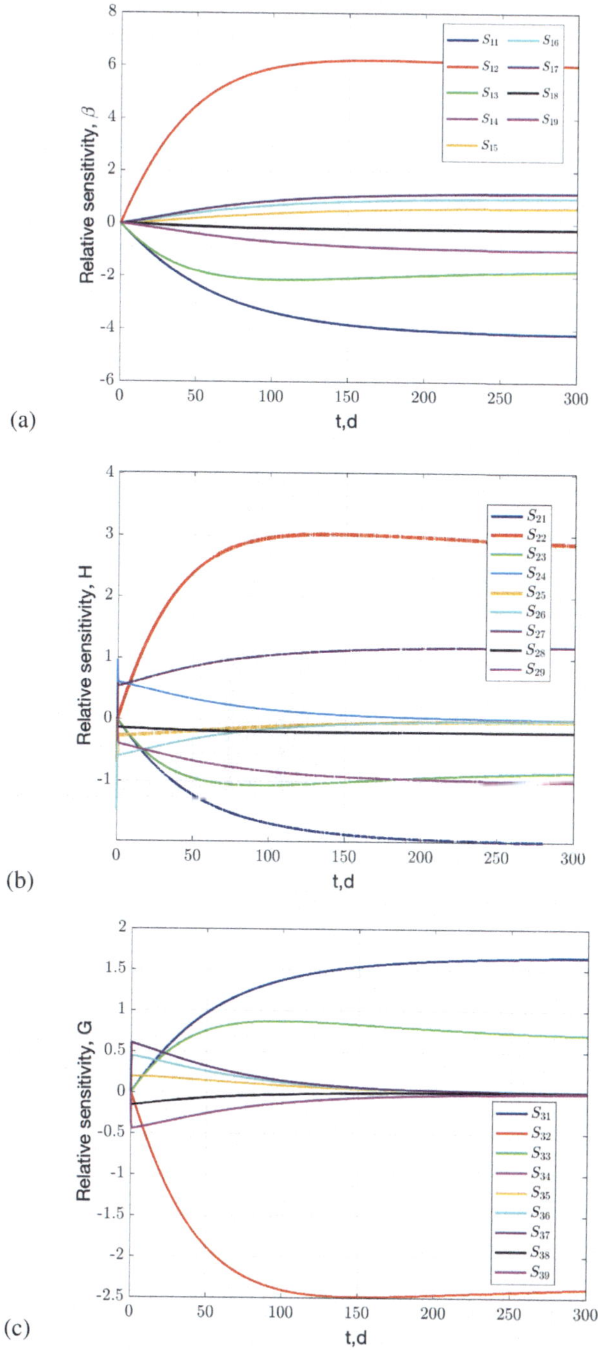

Fig. 2.18 Relative sensitivities of state variables to parameter variations over time. Panels (**a**), (**b**), and (**c**) show the time-dependent relative sensitivities of β-cell mass (β), blood insulin concentration (H), and blood glucose concentration (G), respectively, with respect to each of the nine model parameters. The analysis is based on initial conditions $\beta(0) = 100$ mg, $H(0) = 5$ μU/mL, and $G(0) = 80$ mg/dL

Table 2.6 Sensitivity indices

Parameter p	$\left\| \frac{\partial \beta}{\partial p} \frac{p}{\beta} \right\|_2$	$\left\| \frac{\partial H}{\partial p} \frac{p}{H} \right\|_2$	$\left\| \frac{\partial G}{\partial p} \frac{p}{G} \right\|_2$
d_0	1370	668	549
m_1	2191	1075	881
m_2	733	362	296
σ	302	87	67
γ	187	48	37
κ	302	87	67
ν	371	416	86
η	69	77	19
ζ	302	339	67

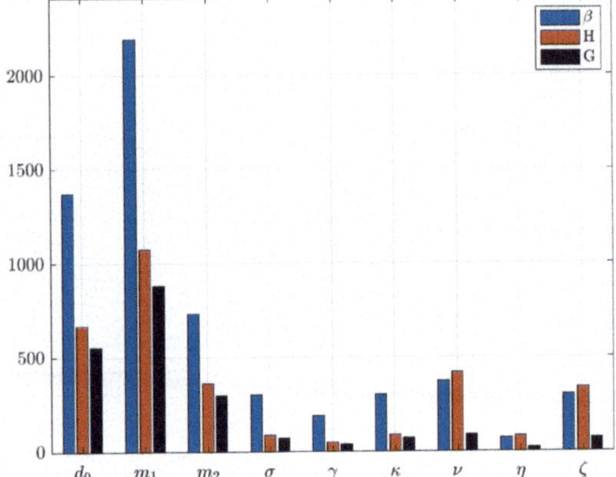

Fig. 2.19 Ranking of sensitivity indices for model parameters. The bar plot shows the relative importance of each of the nine parameters based on their sensitivity indices

2.4.3 Bifurcations and Pathways to Diabetes

A bifurcation occurs when varying a parameter (called a bifurcation parameter) causes a qualitative change in dynamics of a system. Several types of bifurcations are possible (see, e.g., [6, 20, 21]). Among the most common are the saddle-node, transcritical, and Hopf bifurcations. For example, a saddle-node bifurcation occurs when two equilibria approach each other, collide, and annihilate. A transcritical bifurcation involves two equilibria exchanging stability. A Hopf bifurcation occurs when an equilibrium loses stability and gives rise to a stable limit cycle (oscillations).

Having identified parameter m_1 as the most influential through sensitivity analysis, let us explore how variations in m_1 affect the stability of equilibrium solutions. We examine whether changes in m_1 can lead to the disappearance or loss of stability of the physiological equilibrium. This investigation reveals the possibility of bifurcations, qualitative changes in system behavior, that may underlie the onset of diabetes.

Figure 2.20 illustrates the bifurcation diagram, depicting blood glucose concentration G as a function of the parameter m_1, with all other parameters fixed at their values in

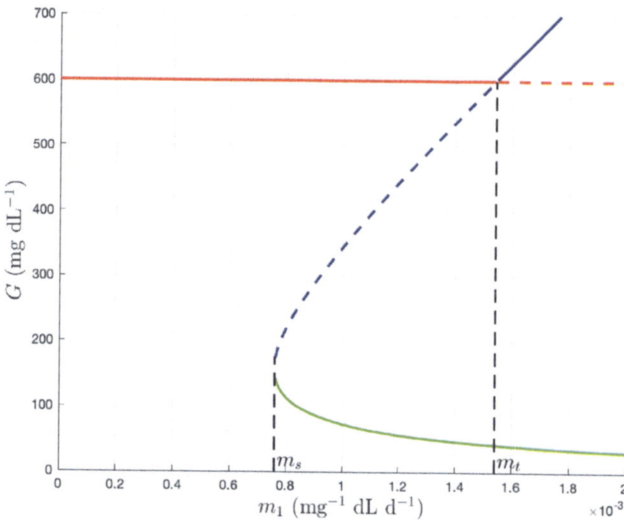

Fig. 2.20 Bifurcation diagram showing blood glucose concentration G as a function of the parameter m_1 with all other parameters fixed at values from Table 2.5. Solid curves indicate stable equilibria; dashed curves indicate unstable equilibria. The red curve represents equilibrium $(\beta_{e0}, H_{e0}, G_{e0})$, the green curve corresponds to equilibrium $(\beta_{e1}, H_{e1}, G_{e1})$, and the blue curve represents equilibrium $(\beta_{e2}, H_{e2}, G_{e2})$. Vertical dashed lines mark the bifurcation points m_s (saddle-node) and m_t (transcritical)

Table 2.5. Two bifurcation values, $m_1 = m_s$ and $m_1 = m_t$, are visible on the diagram. The bifurcation point $m_1 = m_s$ corresponds to a saddle-node bifurcation, while $m_1 = m_t$ corresponds to a transcritical bifurcation. Stable equilibria are shown with solid curves, while unstable equilibria are shown with dashed curves. The red curve represents the pathological equilibrium $(\beta_{e0}, H_{e0}, G_{e0}) = (0, 0, 600)$. The green curve corresponds to the physiological equilibrium $(\beta_{e1}, H_{e1}, G_{e1})$, and the blue curve corresponds to the equilibrium $(\beta_{e2}, H_{e2}, G_{e2})$.

Let us now revisit the equilibrium structure considering this bifurcation behavior. When the parameter m_1 satisfies

$$m_1 > m_s = 2\sqrt{m_2 d_0}, \quad m_1 \neq 2m_2\frac{\nu}{\eta}, \tag{2.82}$$

Equation (2.57) yields two distinct real solutions for the glucose component. These correspond to two equilibria: a stable physiological equilibrium $(\beta_{e1}, H_{e1}, G_{e1}) = (300, 10, 100)$ and an unstable saddle saddle equilibrium $(\beta_{e2}, H_{e2}, G_{e2}) = (37, 2.80, 250)$. Alongside these, a third stable equilibrium exists at $(\beta_{e0}, H_{e0}, G_{e0}) = (0, 0, \frac{\nu}{\eta}) = (0, 0, 600)$, which represents a pathological state characterized by zero β-cell mass and insulin, and elevated glucose levels. Depending on initial conditions, the system may settle into either of the two stable equilibria. As the parameter m_1 decreases and approaches the

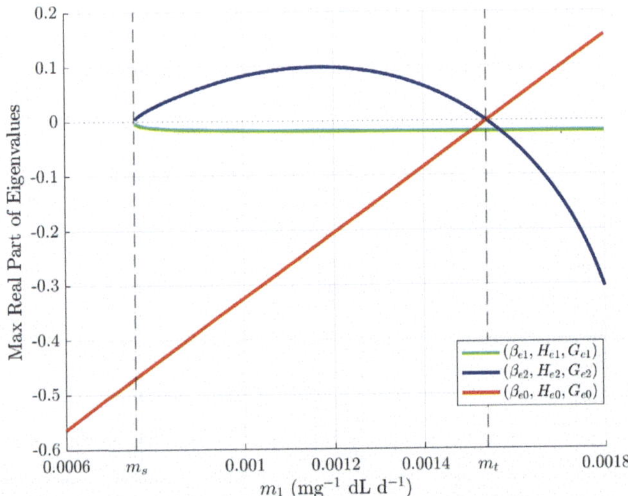

Fig. 2.21 Maximum real part of the eigenvalues of the Jacobian matrix as m_1 varies. The green line shows the equilibrium $(\beta_{e1}, H_{e1}, G_{e1})$, the blue line shows the equilibrium $(\beta_{e2}, H_{e2}, G_{e2})$, and the red line shows the pathological equilibrium $(\beta_{e0}, H_{e0}, G_{e0})$. Vertical dashed lines mark the bifurcation points m_s and m_t, where stability changes occur

bifurcation value $m_s \approx 7.59 \cdot 10^{-4}(\text{mg}^{-1}\ \text{dL}\ d^{-1})$ computed using the parameter values from Table 2.5, the physiological equilibrium (a stable sink) and the saddle equilibrium move closer together. At the bifurcation value $m_1 = m_s$, these two equilibria coalesce into a non-hyperbolic equilibrium $(\beta_{e3}, H_{e3}, G_{e3})$. Evaluating the Jacobian matrix (2.63) at this point reveals a zero eigenvalue, indicating a qualitative change in system behavior typical of a saddle-node bifurcation (see Fig. 2.21 and [16] for details).

When $m_1 < m_s$, the physiological equilibrium disappears, leaving the pathological equilibrium $(\beta_{e0}, H_{e0}, G_{e0}) = (0, 0, 600)$ as the only stable steady state. This change in system behavior offers a mathematical interpretation of diabetes onset as the result of a bifurcation that eliminates the physiological equilibrium.

Next, we consider the second bifurcation parameter:

$$m_t = \frac{d_0\eta^2 + m_2\nu^2}{\nu\eta}, \quad d_0\eta^2 \neq \nu^2 m_2. \tag{2.83}$$

If $m_1 < m_t$, the pathological equilibrium $(\beta_{e0}, H_{e0}, G_{e0}) = \left(0, 0, \frac{\nu}{\eta}\right) = (0, 0, 600)$ remains stable. At the bifurcation point $m_1 = m_t$, the first row of the Jacobian $J(0, 0, \frac{\nu}{\eta})$ becomes a row of zeros, causing the Jacobian matrix (2.63) to have a zero eigenvalue. This signals a transcritical bifurcation where stability is exchanged between equilibria. A more detailed justification of this behavior is also given in [16]; here we just say that if $m_1 > m_t$, the pathological equilibrium $(\beta_{e0}, H_{e0}, G_{e0})$ loses stability, and the saddle equilibrium$(\beta_{e2}, H_{e2}, G_{e2})$ becomes stable (see Figs. 2.20, 2.21). However, this new stable equilibrium has a glucose value $G > 600$ mg $\cdot$ dL^{-1} which is not biologically meaningful because it corresponds to unrealistic negative values of β. Therefore, for $m_1 > m_t$, the system's trajectories are instead attracted to the stable physiological equilibrium $(\beta_{e1}, H_{e1}, G_{e1}) = (300, 10, 100)$.

Using the parameters from Table 2.5, we find that the typical value $m_1 = 0.84 \cdot 10^{-3}(\text{mg}^{-1}\ \text{dL}\ d^{-1})$ is less than $m_t \approx 1.54 \cdot 10^{-3}(\text{mg}^{-1}\ \text{dL}\ d^{-1})$, so under these conditions the pathological equilibrium (0, 0, 600) is stable.

Finally, Hopf bifurcations, which give rise to oscillatory behavior when a pair of complex conjugate eigenvalues crosses the imaginary axis, can also occur in dynamical systems. However, for the parameters used here (from Topp et al. [15]), the characteristic equation (2.67) has only real roots. Thus, no Hopf bifurcations arise in this model, and we do not discuss them further.

The onset of diabetes can be understood as a consequence of how sensitive the system's equilibria are to changes in key parameters. The analysis shows that changes in glucose-insulin dynamics or insulin sensitivity do not necessarily eliminate the physiological equilibrium. However, such changes may increase the required β-cell mass needed to maintain the physiological glucose level. Diabetes may arise as a result of a bifurcation in which the physiological equilibrium loses stability as the parameter m_1 crosses a critical threshold. This transition is illustrated by a bifurcation diagram, and numerical simulations further support the theoretical findings.

In summary, the nonlinear glucose–insulin β-cell model reveals a rich structure of equilibria whose stability depends sensitively on key physiological parameters. Equilibrium analysis identifies both physiological and pathological steady states, while linearization provides insight into their local stability. Sensitivity and bifurcation analyses highlight critical parameters, such as m_1, whose variations can trigger qualitative shifts in system behavior, including the loss of the physiological equilibrium and the onset of diabetes. Together, these analyses demonstrate how mathematical modeling can uncover mechanistic pathways to disease and identify potential targets for intervention.

Exercises

2.1 Using the fasting state conditions described above, derive system (2.2) from the original model (2.1) by introducing the change of variables: $g(t) = G(t) - G_f, \quad h(t) = H(t) - H_f$.

2.2 Derive the expression for $H(t)$ in Eq. (2.7) by solving the nonhomogeneous linear differential equation

$$\frac{dh}{dt} + ch = d\,g(t),$$

with $h(0) = 0$, where $g(t)$ is given by Eq. (2.5).

Note. If you choose the integrating factor method, the integral on the right-hand side will require applying integration by parts twice.

2.3 (Estimating Parameters of a Glucose Response Model) In this exercise, you will fit a mathematical model for blood glucose concentration to observed data. The goal is to estimate the parameters of the function

$$G(t) = G_f + \frac{F}{\omega} e^{-\alpha t} \sin(\omega t),$$

which models the glucose concentration $G(t)$ in the bloodstream following an oral glucose load. The Table below provides the data for blood glucose levels measured at regular time intervals.

1. Fit the model to the data and estimate the parameters F, α, and ω.
   Hint: If needed, you may start with initial guesses such as F=10 $\frac{mg \cdot min}{dL}$, α=0.01 min^{-1}, ω=0.01 min^{-1};
2. Plot the original data and the fitted function on the same graph for visual comparison.
3. Report the fitted parameter values (with units) and interpret their meaning in the context of glucose regulation.

t, min	$G(t)$, mg/dL
0	90
30	150
60	165
90	155
120	135
150	125
180	110
210	100
240	97

2.4 (Insulin Injection and Instantaneous Food Intake) Consider the following Laplace-transformed expressions for two functions $g(t)$ and $h(t)$, involving times t_1 and t_2, and parameters $I_0, K_0, b, d, \alpha, c, \omega$:

$$\mathcal{L}\{g(t)\} = \frac{(s+c)I_0 e^{-st_1} - bK_0 e^{-st_2}}{(s+\alpha)^2 + \omega^2},$$

$$\mathcal{L}\{h(t)\} = \frac{d(s+c)I_0 e^{-st_1} + K_0 e^{-st_2}\left[(s+\alpha)^2 + \omega^2\right] - bd}{(s+c)\left[(s+\alpha)^2 + \omega^2\right]}.$$

Perform analytical inverse Laplace transforms to derive the expressions for $g(t)$ and $h(t)$. Use a symbolic software tool to verify your results. Plot the results.

2.5 (Modeling Food Intake with a Step Function and Insulin Treatment with a Delta Function) Given the Laplace-transformed system derived from the initial value problem (2.8):

$$\begin{cases} \mathcal{L}\{g\}(s+a) &= -b\mathcal{L}\{h\} + I_0\left(\dfrac{e^{-s\tau_1}}{s} - \dfrac{e^{-s\tau_2}}{s}\right), \\ \mathcal{L}\{h\}(s+c) &= d\mathcal{L}\{g\} + K_0 e^{-st_0}, \end{cases}$$

algebraically solve for $\mathcal{L}\{g\}$ and $\mathcal{L}\{h\}$ to derive the expressions given below:

$$\mathcal{L}\{g\} = \frac{(s+c)}{(s+\alpha)^2+\omega^2}\left[I_0\left(\frac{e^{-s\tau_1}-e^{-s\tau_2}}{s}\right) - bK_0\frac{e^{-st_0}}{(s+c)}\right],$$

$$\mathcal{L}\{h\} = \frac{d}{(s+c)}\mathcal{L}\{g\} + \frac{K_0}{(s+c)}e^{-st_0}.$$

2.6 (Modeling Food Intake and Insulin Treatment Using Step Functions) Given the Laplace-transformed system:

$$\begin{cases} \mathcal{L}\{g\}(s+a) &= -b\mathcal{L}\{h\} + I_0\left(\dfrac{e^{-s\tau_1}}{s} - \dfrac{e^{-s\tau_2}}{s}\right) \\ \mathcal{L}\{h\}(s+c) &= d\mathcal{L}\{g\} + K_0\left(\dfrac{e^{-s\lambda_1}}{s} - \dfrac{e^{-s\lambda_2}}{s}\right). \end{cases}$$

algebraically solve for $\mathcal{L}\{g\}$ and $\mathcal{L}\{h\}$ to derive the expressions given below:

$$\mathcal{L}\{g\} = \frac{s+c}{(s+\alpha)^2+\omega^2}\left[I_0\left(\frac{e^{-s\tau_1}-e^{-s\tau_2}}{s}\right) - \frac{bK_0}{(s+c)}\left(\frac{e^{-s\lambda_1}-e^{-s\lambda_2}}{s}\right)\right].$$

$$\mathcal{L}\{h\} = \frac{d}{(s+c)}\mathcal{L}\{g\} + \frac{K_0}{(s+c)}\left(\frac{e^{-s\lambda_1}}{s} - \frac{e^{-s\lambda_2}}{s}\right).$$

2.7 (Modeling Insulin Pill) Consider glucose-insulin dynamics governed by the initial value problem (2.8).

Model the meal as a Dirac delta function at time t_1: $I(t) = I_0\,\delta(t-t_1)$, where I_0 is the amount of glucose ingested.

Represent the insulin pill as a short-term input that begins at time t_2 (time of ingestion) and exhibits absorption followed by elimination:

$$K(t) = K_0\left(e^{-k_e(t-t_2)} - e^{-k_a(t-t_2)}\right)U(t-t_2),$$

where $U(t-t_2)$ is the Heaviside function, K_0 is a scaling factor, k_a is the absorption rate, and k_e is the elimination rate.

(a) Apply the Laplace transform to solve the system in the s domain.
(b) Choose reasonable parameter values and solve the system for $g(t)$ and $h(t)$ numerically in MATLAB. (For example, you may choose $k_a = 0.1 - 0.5\,\text{min}^{-1}$, $k_e = 0.02 - 0.2\,\text{min}^{-1}$, and scale K_0 so that the insulin magnitude is comparable to your fitted values. Explore different pill timings t_0 relative to the meal t_f and analyze how the glucose peak changes).

2.8 Consider the LQR problem defined by the state equations (2.16) and the cost functional (2.17), with the Hamiltonian defined in (2.18). Show that the cost functional can be rewritten as in Eq. (2.19):

$$\mathcal{J} = \frac{1}{2}\int_0^\tau \left[\mathcal{H} - \boldsymbol{\lambda}^T \frac{d\mathbf{x}}{dt}\right] dt.$$

2.9 Consider the Hamiltonian (2.18). Use the optimality condition (2.22) to show that the optimal control is given by Eq. (2.24)

$$\mathbf{u} = -R^{-1}B^T\boldsymbol{\lambda}.$$

2.10 Using the Hamiltonian defined in (2.18), the costate equation (2.21), and the transversality condition (2.23), derive the costate dynamics for a linear system with quadratic cost. Specifically, derive the Eq. (2.25):

$$\frac{d\boldsymbol{\lambda}}{dt} = -Q\mathbf{x} - A^T\boldsymbol{\lambda} + Ql, \quad \boldsymbol{\lambda}(\tau) = \mathbf{0}.$$

2.11 Consider a linear system given by

$$\dot{\mathbf{x}}(t) = A\mathbf{x}(t) + Bu(t), \quad \mathbf{x}(0) = \mathbf{x}_0,$$

with

$$A = \begin{bmatrix} 0 & 1 \\ -1 & -2 \end{bmatrix}, \quad B = \begin{bmatrix} 0 \\ 1 \end{bmatrix}.$$

The cost functional is:

$$\mathcal{J}(u) = \frac{1}{2}\int_0^\tau \left[\mathbf{x}^T(t)Q\mathbf{x}(t) + u^2(t)\right] dt,$$

where $Q = \begin{bmatrix} 1 & 0 \\ 0 & 0 \end{bmatrix}$, and $\tau > 0$ is a given terminal time.

(a) Formulate the Hamiltonian $\mathcal{H}(\mathbf{x}, u, \boldsymbol{\lambda}, t)$.
(b) Derive the adjoint equations and the optimality condition for $u(t)$.
(c) State the boundary conditions for this optimal control problem.

2.12 Consider a linear system given by

$$\begin{cases} \dfrac{dx}{dt} = x + y, \\ \dfrac{dy}{dt} = -2y + u. \end{cases}$$

The cost functional is:

$$\mathcal{J}(u) = \frac{1}{2}\int_0^{\tau} u^2(t)dt,$$

where $\tau > 0$ is a given terminal time.

(a) Formulate the Hamiltonian $\mathcal{H}(\mathbf{x}, u, \boldsymbol{\lambda}, t)$.
(b) Derive the adjoint equations and the optimality condition for $u(t)$.
(c) State the transversality condition.

2.13 Solve the initial value problem given in Eq. (2.39):

$$\begin{cases} \dfrac{d^2g}{dt^2} + (Y_2 + a + c)\dfrac{dg}{dt} - \xi g = -bY, \\ g(0) = g_0, \quad \dfrac{dg}{dt}(0) = -ag_0 - bh_0. \end{cases}$$

(a) Solve the homogeneous part of the equation and find the general solution.
(b) Find a particular (steady-state) solution to the nonhomogeneous equation.
(c) Use the initial conditions to determine the constants of integration.
(d) Verify that the complete solution can be written in the form shown in Eq. (2.40):

$$g(t) = (C_1\cos(\omega t) + C_2\sin(\omega t))\, e^{-\alpha t} + g_p.$$

Determine explicit expressions for C_1, C_2, α, ω, and g_p.

2.14 Using the analytical formulas derived in Eqs. (2.40), (2.43), and (2.44), compute and plot the time evolution of glucose concentration $g(t)$, insulin $h(t)$ and control variable $u(t)$ for the remaining patients listed in Table 2.4. Use the same initial condition $g(0) = 300\,\text{mg/dL}$, target glucose level $l = 100\,\text{mg/dL}$ and weighting parameter $\rho = 10\,\frac{\text{mg}^2\cdot\text{min}^2}{\mu\text{U}^2}$. Compare and interpret the differences in glucose regulation dynamics between patients.

2.15 Choose one patient from Table 2.4 and study how the weighting parameter ρ in the cost functional influences the optimal control.

Using Eqs. (2.40), (2.43), and (2.44), compute and plot $g(t)$, $h(t)$, and $u(t)$ for three different values of ρ. Assume $g(0) = 300\,\text{mg/dL}$, $h(0) = 0\,\mu\text{U/mL}$, and target level $l = 100\,\text{mg/dL}$.

Compare the results. How does changing ρ affect the magnitude of $u(t)$, the speed of glucose regulation, and the behavior of $h(t)$? Discuss the trade-off between glucose control and insulin usage.

2.16 Consider the linear system

$$\frac{d\mathbf{x}}{dt} = M\mathbf{x} + \mathbf{n}, \qquad \mathbf{x} = [g, h, \lambda_1, \lambda_2]^T,$$

with

$$M = \begin{bmatrix} -a & -b & 0 & 0 \\ d & -c & 0 & -\frac{1}{2\rho} \\ -2 & 0 & a & -d \\ 0 & 0 & b & c \end{bmatrix}, \qquad \mathbf{n} = \begin{bmatrix} 0 \\ 0 \\ 2l \\ 0 \end{bmatrix}.$$

Answer the following:

(a) Derive the characteristic equation for the matrix M. Show that, by setting $y = r^2$, the variable y satisfies

$$y^2 - A_1 y + A_2 = 0, \qquad A_1 = a^2 + c^2 - 2bd, \quad A_2 = (ac + bd)^2 + \frac{b^2}{\rho}.$$

Hence, recover the four roots of M in the form

$$r = \pm\sqrt{\frac{A_1 \pm \sqrt{A_1^2 - 4A_2}}{2}}.$$

(b) For each root r_i, solve the homogeneous system

$$(M - r_i I_4)\mathbf{v}_i = \mathbf{0},$$

and show that one may choose the eigenvectors in the form

$$\mathbf{v}_i = \begin{bmatrix} \frac{1}{2}\left(a - r_i + \dfrac{db}{c - r_i}\right) \\ \dfrac{1}{c + r_i}\left(\frac{d}{2}\left(a - r_i + \dfrac{db}{c - r_i}\right) + \dfrac{b}{2\rho(c - r_i)}\right) \\ 1 \\ -\dfrac{b}{c - r_i} \end{bmatrix}.$$

2.17 Consider the homogeneous linear system:

$$\frac{d\mathbf{x}}{dt} = M\mathbf{x}$$

where the matrix $M \in \mathbb{R}^{4\times 4}$ and $\mathbf{x}(t) \in \mathbb{R}^4$.

Suppose the eigenvalues of M are given by:

$$r_{1,2} = \alpha \pm i\omega, \quad r_{3,4} = -\alpha \pm i\omega, \quad \text{with } \alpha, \omega > 0,$$

and that the corresponding eigenvectors appear in complex-conjugate pairs:

$$\mathbf{v}_2 = \mathbf{v}_1^*, \quad \mathbf{v}_4 = \mathbf{v}_3^*$$

(a) Show that the general solution to the system can be written as:

$$\mathbf{x}_h(t) = \sum_{i=1}^{4} C_i \mathbf{v}_i e^{r_i t},$$

where $C_i \in \mathbb{C}$, and $\mathbf{v}_i$ are the eigenvectors corresponding to eigenvalues r_i.

(b) Use the fact that the eigenvalues and eigenvectors come in complex-conjugate pairs to rewrite the solution $\mathbf{x}_h(t)$ in a real-valued form involving trigonometric functions. Derive an expression of the form:

$$\mathbf{x}_h(t) = e^{\alpha t}\left[C^{(1)}\Re(\mathbf{v}_1)\cos(\omega t) + C^{(2)}\Im(\mathbf{v}_1)\sin(\omega t)\right] + e^{-\alpha t}\left[C^{(3)}\Re(\mathbf{v}_3)\cos(\omega t) + C^{(4)}\Im(\mathbf{v}_3)\sin(\omega t)\right],$$

where $C^{(1)}, C^{(2)}, C^{(3)}, C^{(4)} \in \mathbb{R}$.

Hint: Use the identity:

$$\Re\left(\mathbf{v}e^{(\alpha+i\omega)t}\right) = e^{\alpha t}\left[\Re(\mathbf{v})\cos(\omega t) - \Im(\mathbf{v})\sin(\omega t)\right].$$

2.18 Find the constant particular solution $\mathbf{x}_p$ of the system

$$\frac{d\mathbf{x}}{dt} = M\mathbf{x} + \mathbf{n}$$

by solving the algebraic system

$$M\mathbf{x}_p + \mathbf{n} = \mathbf{0}.$$

Show that

$$\mathbf{x}_p = \begin{bmatrix} \dfrac{lb^2}{b^2+\rho(ac+bd)^2} \\ \dfrac{-abl}{b^2+\rho(ac+bd)^2} \\ \dfrac{-2\rho cl(ac+bd)}{b^2+\rho(ac+bd)^2} \\ \dfrac{2\rho bl(ac+bd)}{b^2+\rho(ac+bd)^2} \end{bmatrix}. \tag{2.84}$$

Verify that the first two components of $\mathbf{x}_p$ coincide with the steady-state glucose and insulin values given in (2.40) and (2.43). Briefly interpret these results in the context of glucose-insulin regulation.

2.19 Consider the system (2.47) with two-point boundary conditions:

$$g(0) = g_0 = 280\text{ mg/dL}, \quad h(0) = 0, \quad \lambda_1(\tau) = \lambda_2(\tau) = 0, \quad \tau = 180\text{ min}.$$

Use the parameter values from Table 2.4 (Patient D4a), target glucose level $l = 100\,\text{mg/dL}$, and $\rho = 10\,\frac{\text{mg}^2\cdot\text{min}^2}{\mu\text{U}^2}$.

(a) Compute the homogeneous solution

$$\mathbf{x}_h(t) = \sum_{i=1}^{4} C_i\mathbf{v}_i e^{r_i t}$$

using the eigenvalues r_i and eigenvectors $\mathbf{v}_i$ derived in Exercise 2.16 and the particular solution $\mathbf{x}_p$ from Exercise 2.18.

(b) Determine the constants C_i such that the solution satisfies the given two-point boundary conditions.
(c) Plot the first two components $g(t)$ and $h(t)$ over $t \in [0, \tau]$.
(d) Using the fourth component $\lambda_2(t)$, compute the optimal control

$$u(t) = -\frac{\lambda_2(t)}{2\rho}$$

and plot it over the same interval.
(e) Briefly discuss the qualitative features of the glucose and insulin trajectories and the variation of control effort over time.

2.20 Explain why the Forward–Backward Sweep Method requires both a forward integration and a backward integration. Which equations are solved forward, and which backward?

2.21 Minimize

$$\mathcal{J}(u) = \int_0^1 \left(x(t)^2 + u(t)^2\right) dt$$

subject to

$$\frac{dx(t)}{dt} = -x(t) + u(t), \quad x(0) = 1$$

(a) Write down the Hamiltonian and adjoint equation for this problem.
(b) Perform one forward–backward sweep iteration. Start with an initial guess $u(t) = 0$ and show the updated state, adjoint, and control.
(c) Implement the Forward-Backward Sweep Method with the following relaxation parameters $\theta = 0.2, 0.5, 0.8$ and fixed tolerance $\delta = 10^{-5}$. Compare the number of iterations in each case.
(d) Implement the Forward-Backward Sweep Method with the following tolerances $\delta = 10^{-2}, 10^{-4}, 10^{-6}$ and fixed $\theta = 0.5$. Compare the number of iterations in each case.
(e) Compare your final numerical control $u(t)$ and state $x(t)$ from parts (c) and (d) with the exact analytical solution.

2.22 Repeat the numerical procedure described in Subsection 2.3.4, using the Forward-Backward Sweep Method and the parameters for patient 4a (see Table 2.4). Keep the same initial glucose level $g(0) = g_0 = 300\,\text{mg/dL}$, desired target level $l = 100\,\text{mg/dL}$ and $\rho = 10\frac{\text{mg}^2 min^2}{\mu\text{U}^2}$. Implement the forward-backward sweep method over time interval $[0, \tau]$, where $\tau = 180$ minutes. Plot the resulting trajectories of glucose $g(t)$, insulin $h(t)$, and control $u(t)$.

2.23 Suppose your forward–backward sweep uses tolerance $\delta = 10^{-5}$ with the l^2 norm.

(a) Explain how you check convergence at each iteration.
(b) If the maximum change between consecutive controls is $6.7 \cdot 10^{-6}$, does the iteration stop?

2.24 Consider the following system of differential equations modeling the glucose-insulin-β-cell dynamics (system (2.55)):

$$\begin{cases} \dfrac{d\beta}{dt} = (-d_0 + m_1 G - m_2 G^2)\beta, \\ \dfrac{dH}{dt} = \dfrac{\sigma G^2}{\gamma + G^2}\beta - \kappa H, \\ \dfrac{dG}{dt} = \nu - \eta G - \zeta H G. \end{cases}$$

Show that if the initial conditions are nonnegative, i.e., $\beta(0) \geq 0$, $H(0) \geq 0$, and $G(0) \geq 0$, then the solution remains nonnegative at least locally. In other words, demonstrate that the solution stays in the biologically meaningful region where all variables are nonnegative.

Hint: Check the signs of the right-hand sides when one or more variables are zero.

2.25 Using nullcline analysis, determine the equilibrium solutions of the system from Exercise 2.24 (system (2.55)) under the assumption that the β-cell mass remains constant and nonzero.

(a) Choose several fixed values of β (e.g., $\beta = 50,\ 150,\ 300,\ 450$ mg).
(b) For each case, plot the G-nullcline and the corresponding H-nullcline(s) on the same coordinate plane.
(c) Based on your plots, describe how increasing β-cell mass influences the equilibrium levels of glucose and insulin.

2.26 Solve the system from Exercise 2.24 (system (2.55)) numerically to obtain the time evolution of blood glucose concentration $G(t)$, blood insulin concentration $H(t)$, and β-cell mass $\beta(t)$ for the following sets of initial conditions:

(a) $\beta(0) = 30$ mg, $H(0) = 4\ \mu\text{U} \cdot \text{mL}^{-1}$, $G(0) = 300\ \text{mg} \cdot \text{dL}^{-1}$;
(b) $\beta(0) = 30$ mg, $H(0) = 2\ \mu\text{U} \cdot \text{mL}^{-1}$, $G(0) = 250\ \text{mg} \cdot \text{dL}^{-1}$;
(c) $\beta(0) = 25$ mg, $H(0) = 5\ \mu\text{U} \cdot \text{mL}^{-1}$, $G(0) = 200\ \text{mg} \cdot \text{dL}^{-1}$

Plot the trajectories of all three variables over time for each case. Compare the results and determine whether the solutions approach (converge) or move away (diverge) from the equilibrium point $(\beta_{e0}, H_{e0}, G_{e0})$.

Based on your findings, discuss whether the system's behavior is consistent with our earlier conclusions about the stability of that equilibrium.

2.27 Solve the system from Exercise 2.24 (system (2.55)) numerically to obtain the time evolution of blood glucose concentration $G(t)$, blood insulin concentration $H(t)$, and β-cell mass $\beta(t)$ for the following sets of initial conditions:

(a) $\beta(0) = 100$ mg, $H(0) = 5\ \mu\text{U} \cdot \text{mL}^{-1}$, $G(0) = 80\ \text{mg} \cdot \text{dL}^{-1}$;
(b) $\beta(0) = 90$ mg, $H(0) = 10\ \mu\text{U} \cdot \text{mL}^{-1}$, $G(0) = 300\ \text{mg} \cdot \text{dL}^{-1}$;
(c) $\beta(0) = 40$ mg, $H(0) = 10\ \mu\text{U} \cdot \text{mL}^{-1}$, $G(0) = 200\ \text{mg} \cdot \text{dL}^{-1}$

Plot the trajectories of all three variables over time for each case. Compare the results and determine whether the solutions approach (converge) or move away (diverge) from the equilibrium point $(\beta_{e1}, H_{e1}, G_{e1})$.

Based on your findings, discuss whether the system's behavior is consistent with our earlier conclusions about the stability of that equilibrium.

2.28 Based on existing studies [19], physical exercise can increase insulin sensitivity ζ by up to 36%, while factors such as age and obesity may decrease it by as much as 60%. Investigate the relative sensitivities of β, H, and G to changes in the parameter ζ. In particular, explore whether these changes in ζ can induce a shift in the physiological equilibrium $(\beta_{e1}, H_{e1}, G_{e1})$ of the system.

Use the following scenarios for your analysis:

(a) Increase ζ by 36% from its baseline value in Table 2.5.
(b) Decrease ζ by 60% from its baseline value in Table 2.5.

2.29 (β-Cell Mass Unresponsiveness) Consider a scenario in which β remains constant, indicating that the β-cell mass no longer responds to changes in blood glucose concentration. As a result, the system reduces to the second and third equations of system (2.55):

$$\begin{cases} \dfrac{dH}{dt} &= \frac{\sigma G^2}{\gamma + G^2}\beta - \kappa H, \\ \dfrac{dG}{dt} &= \nu - \eta G - \zeta H G. \end{cases}$$

Apply the direct differential parameter sensitivity method introduced in Sect. 3.4.2 to rank the six parameters in this reduced two-dimensional system. Summarize your findings. In particular:

(a) Are insulin sensitivity ζ and the glucose production rate ν the most influential parameters?
(b) Investigate how different fixed values of β affect the equilibrium glucose concentration. Compare your results with those from Exercise 2.25.

2.30 Give a definition of a bifurcation in a dynamical system. Then describe the qualitative behavior of solutions near each of the following:

(a) A saddle-node bifurcation.
(b) A transcritical bifurcation.

Include sketches if helpful to support your explanation.

2.31 (Bifurcation and Disease Onset) The parameter m_1 plays a key role in maintaining glucose balance. Suppose m_1 gradually decreases and crosses the critical threshold m_s defined in (2.82).

(a) What happens to the number and stability of the system's equilibria as m_1 passes through m_s?
(b) How does this change affect the long-term behavior of glucose, insulin, and β-cell mass?
(c) Based on your understanding, how can this transition be interpreted in terms of a shift from a healthy to a diabetic state?

2.32 Consider the mathematical model of blood glucose regulation given in Exercise 2.24 (system (2.55)). Assume that the parameter m_1 is the most influential in determining the system's behavior.

Using the parameter values from Table 2.5, write a computer script to:

(a) Compute the system's equilibrium points for a range of m_1 values. The range should include the bifurcation values m_s and m_t defined in Eqs. (2.82) and (2.83).
(b) Evaluate the Jacobian matrix at each equilibrium point.
(c) Calculate and record the eigenvalues of the Jacobian matrix.
(d) Determine the local stability of each equilibrium based on the eigenvalues.
(e) Plot the maximum real part of the eigenvalues as a function of m_1, and clearly indicate the bifurcation values m_s and m_t on the plot.

List of Notations for Chap. 2

Symbol	Description	Units
t	Time	min
$G(t)$	Blood glucose concentration at time t	mg/dL
$H(t)$	Net hormone (e.g., insulin) concentration at time t	μU/mL
I_b	Basal glucose input rate	mg/(dL·min)
H_b	Basal hormone secretion rate	μU/(mL·min)
a	Glucose clearance rate	1/min
b	Hormonal effect on glucose clearance	1/(min · μU/mL)
c	Hormone degradation rate	1/min
d	Glucose-induced hormone production rate	1/(min·mg/dL)
$I(t)$	Time-dependent external glucose input (e.g., meals)	mg/(dL·min)
G_f	Fasting blood glucose concentration	mg/dL
H_f	Fasting net hormone concentration	μU/mL
$g(t)$	$G(t) - G_f$	mg/dL
$h(t)$	$H(t) - H_f$	μU/mL
α	Damping coefficient	min^{-1}
ω_0	Natural frequency	min^{-1}
ω	$\omega_0^2 - \alpha^2$	min^{-1}
T	Period of oscillations	min
$\delta(t)$	The Dirac delta function	
s	Parameter of the Laplace transform	
$\mathcal{L}\{g(t)\}$	The Laplace transform of $g(t)$	
$K(t)$	Insulin delivery rate predefined as a bolus or a step input	μU/(mL·min)
$U_\tau(t)$	$U_\tau(t) = \begin{cases} 0, & \text{for } t < \tau, \\ 1, & \text{for } t \geq \tau. \end{cases}$	
l	The desired constant glucose level	mg/dL
$u(t)$	Continuous control input (insulin infusion rate) in optimal control	μU/(mL·min)
$\mathcal{J}$	Cost function in optimal control	$\mu U^2/(mL^2 \cdot min^2)$
ρ	Weighting parameter in optimal control	$(mg^2 \cdot min^2)/\mu U^2$
$\mathbf{x}(t)$	State $\mathbb{R}^n$ vector	
$\mathbf{x}_0$	$\mathbf{x}(t = 0)$	
$A(t)$	State transition $\mathbb{R}^{n\times n}$ matrix	
$B(t)$	Control input $\mathbb{R}^{n\times m}$ matrix	
$Q(t)$	State weighting $\mathbb{R}^{n\times n}$ matrix in cost functional	
$R(t)$	Control weighting $\mathbb{R}^{m\times m}$ matrix in cost functional	

List of Notations for Chap. 2 (Continues)

Symbol	Description	Units
l	Target state $\mathbb{R}^n$ vector	
$\mathbf{e}(t)$	Deviation from target, $\mathbf{x}(t) - \mathbf{l}$	
$\boldsymbol{\lambda}(t)$	Adjoint (costate) $\mathbb{R}^n$ vector	
$\mathcal{H}$	Hamiltonian	
$\boldsymbol{P}(t)$	Riccati $\mathbb{R}^{n\times n}$ matrix in expression (2.26). In 2×2 case, $P = \begin{bmatrix} q & z \\ z & j \end{bmatrix}, q, z, j \in \mathbb{R}$	
$\boldsymbol{\mu}(t)$	Inhomogeneous term ($\mathbb{R}^n$ vector) in expression (2.26)	
Y_0	$\frac{q}{2\rho}$	
Y_1	$\frac{z}{2\rho}$	
Y_2	$\frac{j}{2\rho}$	
y	$Y_2^2 + 2(a+c)Y_2$	
ξ	$-\sqrt{(ac+bd)^2 + \frac{b^2}{\rho}}$	
Y	$\frac{bl}{\rho\xi}$	
g_p	Particular solution of (2.40)	
C_1, C_2	Integration constants in $g(t)$	μU/mL
D_1, D_2	Integration constants in $h(t)$	mg/dL
E_1, E_2	Integration constants in $u(t)$	μU/(mL·min)
E_3	$(-Y_1 + \frac{a}{b}Y_2)g_p + Y$	
M	Coefficient $\mathbb{R}^{4\times 4}$ matrix in linear system (2.47)	
$\mathbf{n}$	Constant source $\mathbb{R}^4$ vector in (2.47)	
$r_i,\ i = 1, 2, \ldots, n$	Eigenvalues of a $(n\times n)$ matrix	
$\boldsymbol{v}_i,\ i = 1, 2, \ldots, n$	Eigenvectors of a $(n\times n)$ matrix	
$C^{(i)},\ i = 1, 2, 3, 4$	Integration constants in (2.52)	
θ	relaxation parameter in the Forward-Backward Sweep Method	
δ	convergence tolerance in the Forward-Backward Sweep Method	
$\beta(t)$	β-cell mass	mg
d_0	Death rate of β-cells at zero glucose	d^{-1}
m_1	Coefficient for glucose-stimulated replication of β-cells	$(mg\cdot d)^{-1}dL$
m_2	Coefficient for glucose-induced death of β-cells	$mg^{-2}dL^2d^{-1}$
σ	Maximal insulin secretion rate per β-cell	$\mu U(mg\cdot mL\cdot d)^{-1}$
γ	Glucose level at which secretion rate reaches half-maximum	mg^2dL^{-2}

List of Notations for Chap. 2 (Continues)

Symbol	Description	Units
κ	Insulin removal rate constant	d^{-1}
ν	Constant glucose production rate	$mg(dL \cdot d)^{-1}$
η	Coefficient for insulin-independent glucose removal	d^{-1}
ζ	Coefficient for insulin sensitivity	$mL(\mu U \cdot d)^{-1}$
(β_e, H_e, G_e)	Equilibrium solutions of system (2.55)	
$\boldsymbol{p}$	$\mathbb{R}^9$ vector of parameters (see Table 2.5)	
J	Jacobian	
$\boldsymbol{s}_j$	Sensitivity $\mathbb{R}^3$ vector for parameter $p_j,\ j = 1, 2, \ldots, 9$	
S	Sensitivity $\mathbb{R}^{3\times 9}$ matrix	
$\boldsymbol{f}$	$\mathbb{R}^{3\times 9}$ matrix in (2.77)	
$S_{i,j}$	Relative sensitivity $\frac{\partial x_i}{\partial p_j} \frac{p_j}{x_i}$	
$\lVert S_{ij} \rVert_2$	Sensitivity index (see (2.80))	

References

1. Center for disease control and prevention, National Diabetes Statistics Report,n/a https://www.cdc.gov/diabetes/php/data-research/index.html, accessed May 6, 2025.
2. Center for disease control and prevention, diabetes basics,n/a https://www.cdc.gov/diabetes/about/index.html, accessed May 6, 2025.
3. Shahid, R. K., Ahmed, S., Le, D., Yadav, S.: Diabetes and Cancer: Risk, Challenges, Management and Outcomes. Cancers, 13(22), 5735.(2021) https://doi.org/10.3390/cancers13225735
4. Ackerman, E., Gatewood, L.C., Rosevear, J.W., Molnar, G.D.: Model studies of blood-glucose regulation, Buletin of Mathematical Biophysics, 27(1965), pp. 21–37.
5. Ackerman, E., Rosevear, J. W., McGuckin, W.F.: A mathematical model of the Glucose-tolerance test, Phys. Med. Biol. 9(2), pp.203–213.(1964). https://doi.org/10.1088/0031-9155/9/2/307
6. Blanchard, P., Devaney, R. L., Hall, G. R.: Differential Equations, Brooks/Cole, Cengage, Boston, 2011.
7. The MathWorks, Inc. (2024). MATLAB (Version R2024a) [Computer software]. https://www.mathworks.com, accessed May 6, 2025.
8. Savatorova, V., Panayotova, I., Hallare, M. (2025). Mathematical-modeling-in-Life-Sciences-A-Practical-Project-Based-Approach [Source code]. GitHub. https://github.com/ViktoriaLS/Mathematical-modeling-in-Life-Sciences-A-Practical-Project-Based-Approach
9. Kirk, D.E.: Optimal Control Theory. An Introduction. Dover Publications, Inc.,Mineola, N.Y. (2004)
10. Pontryagin, L.S., Boltyanskii, V.G., Gamkrelidze, R.V., Mishchenko, E.F.: The Mathematical Theory of Optimal Processes. Wiley New York (1967)

11. Swan, G.W.: Applications of Optimal Control Theory in Biomedicine. Marcel Dekker, Inc., New York and Basel (1984)
12. Yipintsoi, T., Gatewood, L.C. Ackerman, E., Spivak, P.L., Molnar G.D., Rosevear, J.W., Service, F.J.: Mathematical Analysis of Blood Glucose and Plasma Insulin Responses to Insulin Infusion in Healthy and Diabetic Subjects. Comput. Biol. Med. **3**, 71–78 (1973)
13. Lenhart, S., Workman, J.T.: Optimal Control Applied to Biological Models. Mathematical and Computational Biology Series, Chapman and Hall/CRC, London (2007)
14. Kharab, A., Guenther, R.B.: An Introduction to Numerical Methods: a MATLAB Approach. Chapman and Hall/CRC, New York (2023)
15. Topp, B., Promislov, K., deVries, G., Muira, R.M.,Finegood, D.T.: A Model of β-Cell Mass, Insulin, and Glucose Kinetics: Pathways to Diabetes, J. Theor. Biol. 206, pp.605–619.(2000).
16. Savatorova, V. L., Talonov, A. V.: Differential equations for modeling pathways leading to diabetes onset, CODEE Journal. 18(1), pp.605–619.(2024) https://doi.org/10.5642/codee.NZHZ5522
17. Saltelli, A.: Sensitivity Analysis. John Willey and Sons, Ltd. (2000)
18. Zi, Z.: Sensitivity analysis approaches applied to systems biology models. IET systems biology 5(6), pp.336–346. (2011).
19. Hernandez, R.D., Lyles, D.J., Rubin, D.B., Voden, T.B. and Wirkus, S.A.: A Model of Beta-cell Mass, Insulin, Glucose and Receptor Dynamics with Applications to Diabetes. *Technical Report*, Biometric Department, MTBI Cornell University. (2001) https://ecommons.cornell.edu/items/ae24dc67-d5cd-489a-af70-fb4c48d64408, accessed August 12, 2025.
20. Hirsch, M. W., Smale, S., Devaney, R.L.: Differential Equations, Dynamical Systems, and an Introduction to Chaos. Elsevier. (2013).
21. Strogatz, S.: Nonlinear Dynamics and Chaos with Applications to Physics, Biology, Chemistry, and Engineering. Westview press. (2015)

Multi-species Modeling in the Chesapeake Bay Ecosystem

3

3.1 Introduction

Chesapeake Bay, the largest estuary in the United States, supports numerous ecologically and economically valuable species. Fisheries in the Chesapeake Bay are vital renewable resources that not only support the economic prosperity of local communities but also play a crucial role in maintaining the biodiversity and ecological balance of the Bay [1]. But the use of these biological resources should be done in a sustainable way as over-exploitation may lead on a long run to the stock depletion which in turn may threaten biodiversity of the ecosystem. The sustainable management of these resources is essential to ensure their availability for future generations. Protecting the Bay's fisheries is therefore key to promoting both ecological health and long-term regional sustainability [2].

To this end, mathematical models provide a powerful framework for exploring fish population dynamics by enabling the systematic testing of different assumptions through variation of model parameters. Such models can generate informed hypotheses for empirical validation, identify key drivers of population change, and reveal potential strategies for sustainable management—much like their use in clinical contexts to narrow treatment possibilities. By capturing mechanisms that cannot be easily addressed through field or laboratory studies alone, modeling deepens our understanding of the biological, ecological, and economic processes shaping fisheries. This approach not only enhances scientific insight but also supports evidence-based decision-making for climate-resilient and optimally harvested fish populations.

In this chapter we show how mathematical modeling can be used to study the population dynamics and harvesting management of fishes. We present inquiry-based modeling projects and case studies designed for use in courses such as Mathematical Modeling, Dynamical Systems, or Advanced Differential Equations. Each project or case study guides the reader through the model-building process, outlining key assumptions and

I. Panayotova et al., *Mathematical Modeling in Life Sciences*, Synthesis Lectures on Mathematics & Statistics, https://doi.org/10.1007/978-3-032-17486-4_3

demonstrating how models are formulated, refined, and applied. Through these projects, students gain an appreciation for the role of mathematical models in analyzing ecological systems, projecting long-term trends, and informing sustainable fisheries management.

In Sect. 3.2 we begin with a single-species logistic growth model, applying it to laboratory-derived observational data. Next, in Sect. 3.3 the classical Lotka–Volterra predator–prey system is introduced and fitted both to laboratory predator–prey data and to field data on Atlantic menhaden and striped bass. Section 3.4 discusses parameter sensitivity analysis, providing a framework to evaluate model robustness by examining how small changes in system parameters influence solution behavior over time. Building on this foundation, a three-species model is developed in Sect. 3.5 to study the biological interactions among three important Chesapeake Bay fishes: Atlantic menhaden, native striped bass, and invasive blue catfish. The model is then used in Sect. 3.6 to study key ecological challenges, including the overfishing of menhaden and the invasiveness of blue catfish as well as to test harvesting strategies aimed at controlling blue catfish populations. The analysis draws on algebraic and calculus-based tools such as solving polynomial equations, computing eigenvalues and eigenvectors, and advanced techniques including the Routh–Hurwitz criteria and the Hartman–Grobman theorem. Finally, in Sect. 3.7, numerical simulations are carried out in MATLAB [3] providing graphical results and enabling the exploration of long-term fish population dynamics under varying ecological and management scenarios.

3.2 Modeling Population Growth with the Logistic Model

Let us begin by modeling the growth of a single-species population. We analyze data for *Paramecium Caudatum*, a unicellular protist of the *Phylum Ciliophora*, shown in Table 3.1. This species is widely distributed across marine, brackish, and freshwater environments. The data come from Gause's classic experiments, documented in his monograph [6], in which he investigated the population dynamics of *Paramecium caudatum* grown in monoculture under controlled laboratory conditions. The counts were estimated microscopically by sampling small volumes, so the units represent relative abundance (number of individuals per unit volume) rather than absolute counts.

We use these laboratory data, rather than field data from fish populations, because they exhibit growth patterns that closely follow the "expected" logistic model behavior. This controlled setting provides a clear and instructive foundation for introducing students

Table 3.1 Population growth data for *Paramecium caudatum* grown in a lab as a monoculture. Population is measured in relative abundance (number of individuals per unit volume). Data are extracted from the gauseR package in R [4, 5]

Day	0	1	2	3	4	5	6	7	8	9	10
Population	1	1.8	3	4.2	5.3	6	6.4	6.5	6.5	6.4	6.3

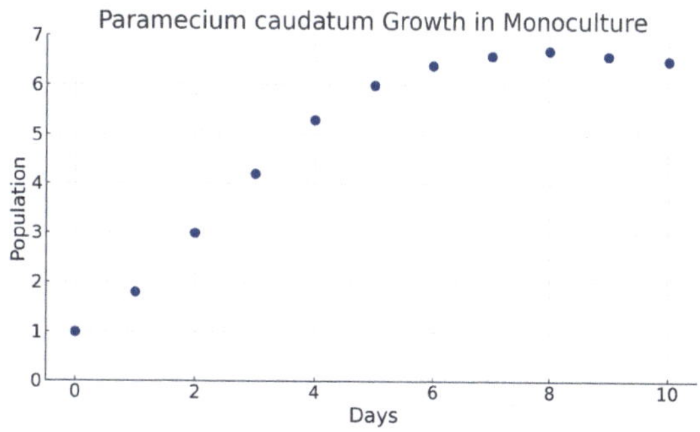

Fig. 3.1 Population growth data of Paramecium caudatum grown as a monoculture. Population is measured in relative abundance (number of individuals per unit volume)

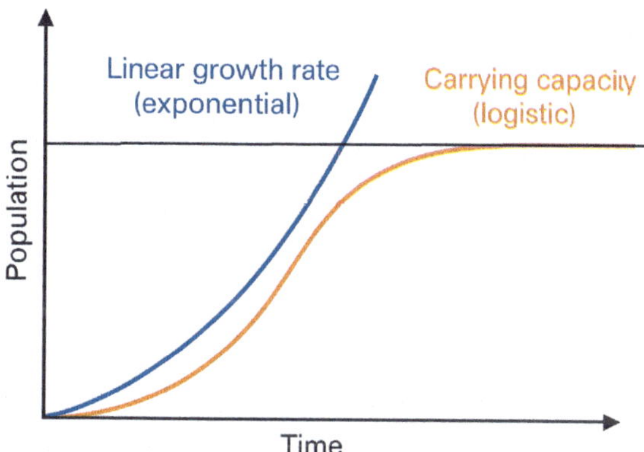

Fig. 3.2 Comparison of exponential and logistic growth models. Incorporating environmental constraints transforms linear (blue curve) growth rate into nonlinear (orange curve) dynamics

to the modeling process before addressing the additional complexity and variability characteristic of natural ecological systems.

From the plotted data in Fig. 3.1 we see that initially, the population increases rapidly, resembling exponential growth. However, after a certain point, the growth rate slows and the population levels off, approaching a stable equilibrium. This plateau phase suggests the presence of limiting factors such as resource or space constraints, even though some natural variability remains in the population size. Such an S-shaped pattern is characteristic of *logistic growth*, where environmental constraints limit further population increase over time.

Figure 3.2 compares the growth predicted by the **exponential** and **logistic models**. In the exponential model, the population increases without bound because resources are assumed to be unlimited. In contrast, the logistic model incorporates density dependence through the carrying capacity, which limits growth as resources become scarce. The population initially grows exponentially but slows as it approaches the carrying capacity, ultimately stabilizing at equilibrium. This comparison illustrates how modifying a single modeling assumption—**resource limitation**—changes both the *quantitative* and *qualitative* behavior of the system.

Mathematically, the exponential growth is modeled by the differential equation $\frac{dx}{dt} = rx$, where the rate of change is assumed to be *linear*, or *proportional to its current size*. Note that **per-capita** growth rate $\frac{dx}{xdt}$ is assumed **constant**. To reflect the fact that resources are limited, the **per-capita growth rate is assumed to decrease linearly with the population size,** $\frac{dx}{xdt} = r - sx$, eventually reaching zero when the population attains its maximum sustainable size, $K = r/s$, known as the carrying capacity K. This leads to the logistic model:

$$\frac{dx}{dt} = rx\left(1 - \frac{x}{K}\right) = x(r - sx), \tag{3.1}$$

where r is the intrinsic growth rate (the net rate at which new individuals are introduced to the population when the population is sparse), s is the density dependence parameter (which reflects how the size of the population affects the overall rate), and K is the carrying capacity (the maximum population size) [7]. Biologically, r captures the reproductive potential of the species under ideal conditions, while K reflects environmental limitations such as available space, food, or nutrients. This framework provides a simple yet powerful model for understanding how populations grow and stabilize over time.

By fitting the model (3.1) to the *Paramecium* data, from Table 3.1 we can estimate biologically meaningful parameters such as the growth rate r and the carrying capacity K, and better understand the population dynamics observed in the experiment.

To perform the fitting we use per-capita growth rate equation in the following form:

$$\frac{dx}{xdt} = r - sx, \tag{3.2}$$

where the right-hand side is an equation of a line. This form is particularly useful when analyzing empirical data as the parameters can be estimated using ordinary least squares regression of per-capita growth rates against species abundances. Hence, if we calculate the per-capita growth rate using the data, we can fit a regression line to those data using MATLAB's function *fitlm*. To find the per-capita growth rate data, we will use the difference equations $dx = x_i - x_{i-1}$ and $dt = t_i - t_{i-1}$, where $i = \overline{1, 10}$. All steps of the model fitting procedure, including data processing, regression analysis, logistic model simulation, and visualization, were performed in MATLAB. The complete code used for the analysis (*Logistic model fitting.mlx*) is provided in GitHub [8] and the fitted model vs observed data are shown in Fig. 3.3.

Note that we can use the model to make some predictions for the population's size for the days after the last day of observation (day 10). As the figure shows, the population approaches a constant equilibrium state which is also the carrying capacity $K = 6.3$ of these population. To quantify how good is the fit of the found model to the given data we calculate the **root mean squared error** (RMSE) defined by the formula:

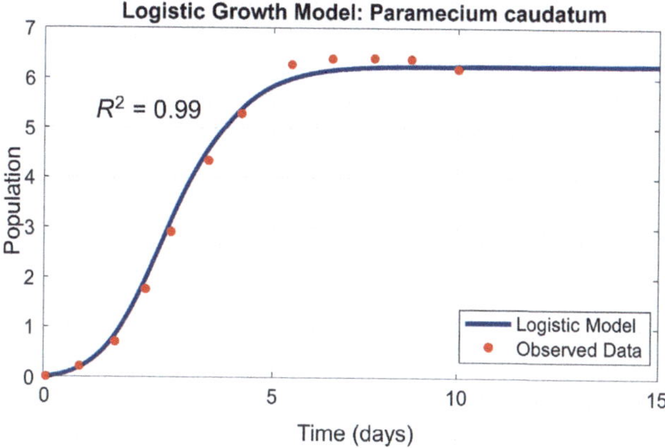

Fig. 3.3 Logistic growth model fitted to *Paramecium caudatum* population data. The blue curve represents the predicted population trajectory based on the model. Red dots show the observed population sizes over time in monoculture

$$RMSE = \sqrt{\frac{1}{n}\sum_{i=1}^{n}(y_i - \hat{y}_i)^2}, \tag{3.3}$$

where y_i and $\hat{y}_i$ represent the observed and predicted (using the model) outputs respectively, and n is the number of observations. This value is a common metric of the accuracy of a model's predictions compared to the given data. It is used to quantify how far off on average the model is from the data. Calculating the RMSE for our model, we get the value $RMSE = 0.0375$ which means that the model's predictions deviate from the observed population size by about 0.04 units, on average, per point. Another common metrics used to assess goodness of fit is the R^2, a statistical measure that tells **how well the model explains the variability** in the observed data.

Definition 3.1 The **goodness of fit measure** R^2 is defined as

$$R^2 = 1 - \frac{SS_{res}}{SS_{tot}}, \quad \text{where} \tag{3.4}$$

- SS_{tot} = total sum of squares = $\sum(y_i - \bar{y})^2$, represents the total variability in the data around the sample mean $\bar{y}$. It shows how "spread out" the observations are, before we use any model.
- SS_{res} = sum of squares of residuals = $\sum(y_i - \hat{y}_i)^2$, represents the leftover variability after the model makes its predictions.

Values of R^2 close to 1 indicate a good fit, whereas values close to zero indicate a poor fit. The logistic model yielded an R^2 value of 0.986, indicating that about 99% of the variability in *Paramecium caudatum* population growth is explained by the model. This suggests a strong fit between the predicted and observed values.

3.3 Modeling Predator-Prey Interactions

In the 1920s, Italian zoologist Umberto D'Ancona (1896–1964) was analyzing data on fish species sold in the markets of the Veneto region of Italy. He noticed a striking trend: following World War I, when commercial fishing in the Adriatic Sea had sharply declined, the proportion of sharks and other predatory fish in market catches increased significantly, even though overall fish landings were low.

D'Ancona hypothesized that this increase in predators was due to the reduced fishing pressure during the war years. With many fishermen enlisted in the military, fewer fish were caught, allowing fish populations to recover and mature. Prey species became more abundant, which in turn supported the growth of predator populations [9].

To explore this hypothesis more rigorously, D'Ancona turned to his father-in-law, the mathematician Vito Volterra. He asked whether a mathematical model could explain why a reduction in fishing might benefit predator species more than prey. Volterra responded by developing a set of differential equations to describe the interaction between two biological populations: predators and their prey.

These equations, now famously known as the Lotka-Volterra equations (named also for Alfred Lotka, who developed a similar model independently [10]), became foundational in theoretical ecology. Volterra modeled the populations of prey, $x(t)$, and predators, $y(t)$, over time t, using the following system of differential equations:

$$\frac{dx}{dt} = rx - bxy, \quad \frac{dy}{dt} = -ey + dxy. \tag{3.5}$$

Here, r, b, e, and d are positive constants representing biological rates [11].

In formulating these equations, Volterra made the following assumptions:

1. **Prey Growth**: The prey population has an abundant food source (such as plankton), so it experiences little competition and can *grow exponentially in the absence of predators*. This explains the term rx in the system (3.5).
2. **Predator Decline**: Predators rely solely on prey for food. Without prey, *the predator population declines exponentially*, represented by the term $-ey$ in (3.5).
3. **Predator-prey interaction:** The interaction between the two populations is modeled through the terms: $-bxy$ and $+dxy$ in (3.5). These terms represent encounters between predators and prey, which are *assumed to occur at a rate proportional to the product of their populations*. Such interactions reduce the prey population (hence the negative sign

in the first term) and support the growth of the predator population (hence the positive sign of the second).

One of the most useful application of a mathematical model is its **predictive power**. In the case of the Lotka-Volterra model, we can use it to simulate and understand the dynamic behaviors of predator-prey systems over time and hence to predict the population dynamics of the two species in future years. Next, we will demonstrate the use of this model with real biological data.

3.3.1 Fitting Lotka-Volterra Predator-Prey Model to Experimental Laboratory Data

In this section we analyze experimental data given in Table 3.2 involving the interaction between two unicellular organisms, *Paramecium aurelia* (prey) and *Didinium nasutum* (predator), grown together under controlled laboratory conditions [6].

When self-limitation is absent for both species, Lotka-Volterra predator-prey model displays *neutrally stable* oscillations, meaning that oscillations maintain a fixed amplitude and period. When self-limitation exists for either the prey or the predator species, or both, then oscillations become damped, meaning that over time they decrease in amplitude and period until they reach stable equilibrium values [7].

Plotting the data from Table 3.2, as given in Fig. 3.4, we observe that the oscillations for both species are damped. Hence, to fit the appropriate model, we need to use the **generalized Lotka-Volterra** model where the prey grows according to the logistic growth in the absence of predator, or in other words, a self-limitation term is included for both species (ax^2 and cy^2):

$$\frac{dx}{dt} = x(r + ax) + bxy, \quad \frac{dy}{dt} = y(e + cy) + dxy. \tag{3.6}$$

Next, we use the open-source software *R* [12, 13] and *gauseR* [4] package to fit the generalized Lotka-Volterra model to this observational data. Fitting generalized Lotka-

Table 3.2 Population growth data for *Paramecium aurelia* (prey) and *Didinium nasutum* (predator), grown together under controlled laboratory conditions. Population is measured in relative abundance (number of individuals per unit volume). Data are extracted from the *gauseR* package in *R* [4, 5]

Day	1	2	3	4	5	6	7	8	9	10
Prey	2.9	7.6	15.2	25.1	55.4	27.1	9.6	2.4	0.5	1.3
Predator	0	0	02	4.8	6.9	31	27.9	15.8	11.7	4.1
Day	11	12	13	14	15	16	17			
Prey	5.3	15.2	27.5	18.2	0	0	0			
Predator	0	0	3.1	28.6	37	18.6	9.3			

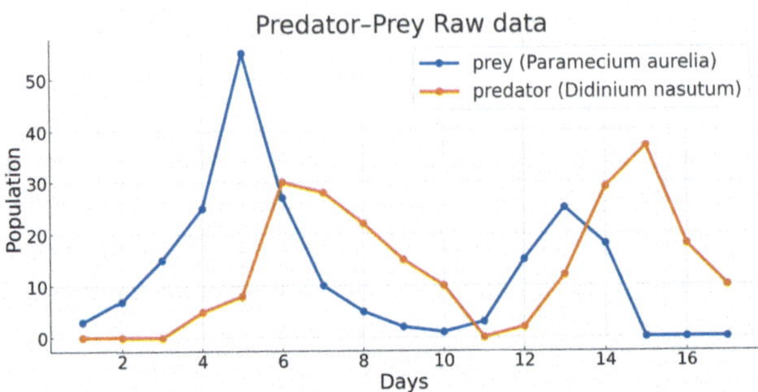

Fig. 3.4 Experimental data of two unicellular organisms, Paramecium aurelia (prey) and Didinium nasutum (predator), grown together under controlled laboratory conditions from [6]. Population is measured in relative abundance (number of individuals per unit volume)

Volterra models in *R* —particularly using the *gauseR* package—tends to work better than in MATLAB due to *R*'s specialized tools for ecological modeling. The *gauseR* package introduced in [5] is specifically optimized for predator-prey dynamics, combining log-space regression with robust ODE-based optimization, and includes safeguards against numerical instability. The package uses solvers from the *deSolve* library, which are well-suited for stiff biological systems and allow for accurate simulation of population dynamics. In contrast, MATLAB's general-purpose solvers and optimizers often struggle with stiff systems and require manual tuning to prevent divergence or biologically unrealistic behavior, making *R* the more reliable choice for this application.

Using the ideas from Sect. 3.2, we divide both sides of Eqs. (3.6) by x and y, respectively, to calculate the **per-capita growth rate** of each species. As a result, the model can be written as a system of two differential equations with right-hand sides that are linear in x and y:

$$\frac{dx}{xdt} = r + ax + by, \quad \frac{dy}{ydt} = e + cy + dx. \tag{3.7}$$

The implementation of the fitting procedure is based on [5] and the code is available on GitHub [8]. Note that in the program the coefficients are named as follows: $r_1 = r, r_2 = e$, $a_{11} = a$, $a_{12} = b$, $a_{21} = d$, and $a_{22} = c$. Next we outline some important steps in the fitting process and explain reasons behind these steps.

The program starts with the calculation of the time-lagged abundances. Next, the least squares linear regression command (*lm*) is used against species abundances to identify the parameters of the mathematical model. If we graph the fitted model against the given data points as shown in Fig. 3.5, we observe that the model does not provide a good fit. This mismatch arises because the parameter estimation method used here is subject to high error, and the Lotka–Volterra system is extremely sensitive—even very small changes in parameter values can lead to large differences in the model's predictions.

To address this issue, we need an optimization approach that directly fits the predicted dynamics to the observed data. In *R*, this can be accomplished using the *optim* command

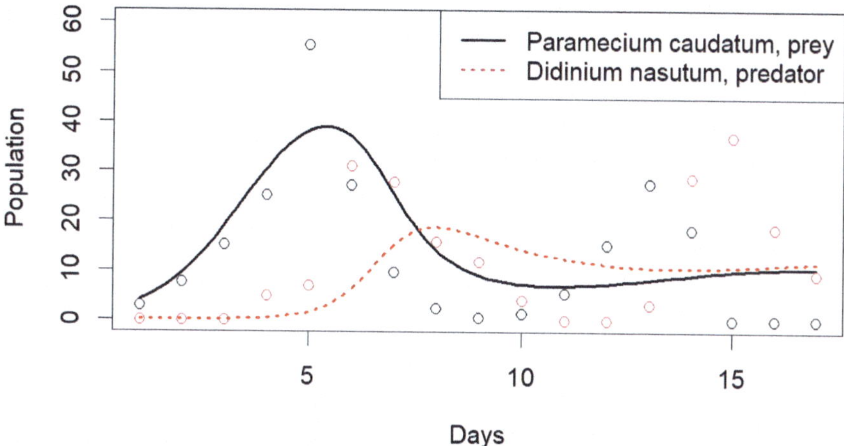

Fig. 3.5 Lotka-Volterra model fitted to the data of Paramecium aurelia (prey) and Didinium nasutum (predator) without optimization

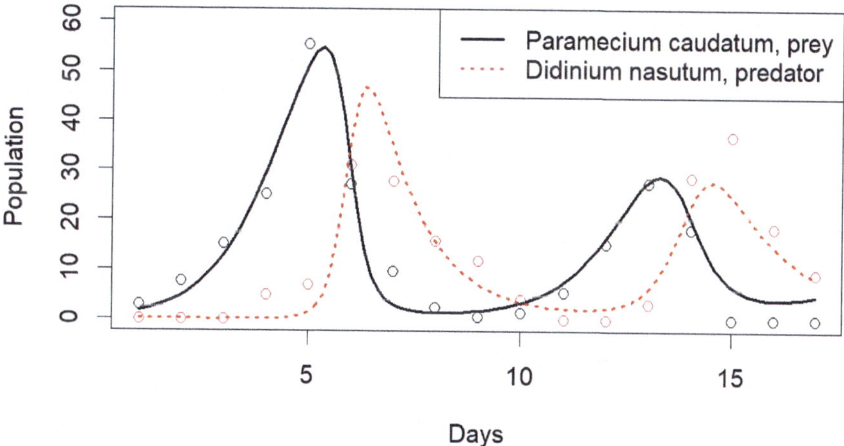

Fig. 3.6 Lotka-Volterra model fitted to the data of *Paramecium aurelia* (prey) and *Didinium nasutum* (predator) with optimization

that minimizes the sum of the squared errors in combination with the *lv-optim* function from the *gauseR* package. Since this optimizer works in log-space, the sign of each parameter (positive or negative) must be recorded and preserved prior to analysis. Once the optimized parameters are obtained, they can be used to define the Lotka–Volterra system of two differential equations, which is then solved numerically with the *deSolve* package using the *ode* command.

The observed data and the fitted model are plotted together in Fig. 3.6 to allow for direct comparison. The improvement in fit demonstrates how optimization reduces the impact of noisy parameter estimates. Because the Lotka–Volterra equations are highly sensitive, even small inaccuracies in parameter values can cause large deviations in model predictions, making this approach especially valuable.

Important: Linear Regression vs Nonlinear Optimization

Linear Regression Fit

- Uses the `lm` command in *R*.
- Provides a quick first approximation by fitting slopes in the per-capita growth equations.
- Captures the broad trends in both prey and predator populations.

Limitations: regression treats the system as linear, so it does not fully respect the nonlinear feedback between predator and prey. Coefficients can sometimes be counterintuitive (e.g., negative prey effects).

Nonlinear Optimization Fit

- Uses the `optim` command in *R*.
- Directly minimizes the mismatch between model trajectories and observed data.
- Produces parameter values that are more consistent with the nonlinear structure of the Lotka–Volterra equations.

3.3.2 Fitting Lotka-Volterra Predator-Prey Model to Fish Field Data

In this section, we analyze field observational data for Atlantic menhaden and striped bass. Our dataset spans over years 1982–2017 and consists of population abundance estimates extracted from the Atlantic States Marine Fisheries Commission's (ASMFC) 2019 Benchmark Stock Assessment for Atlantic menhaden and the 2022 Stock Assessment Update for striped bass [14] which are freely accessible for everyone. To approximate pre-harvesting population levels, we augmented the reported abundance indices by adding estimated recreational and commercial harvest for each species. In other words, we constructed an adjusted abundance index by adding estimated recreational and commercial harvest (removals) to the reported abundance, yielding an approximate measure of total numbers prior to harvesting. The compiled time series file `MenhBassDataInBillions.csv` gives the fish population abundances in billions for both species and could be downloaded from GitHub [8] for analysis.

Plotting the time series data as shown in Fig. 3.7 reveals fluctuations consistent with predator–prey dynamics between Atlantic menhaden (prey) and striped bass (predator). Although the oscillations are less pronounced than in the laboratory experiments discussed

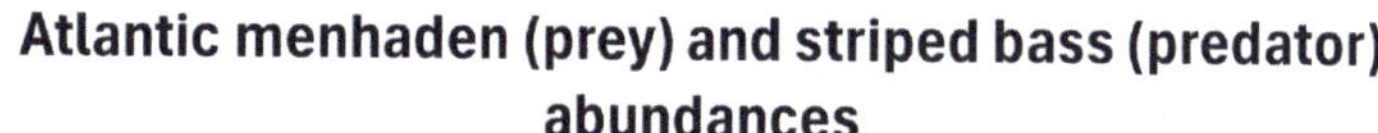

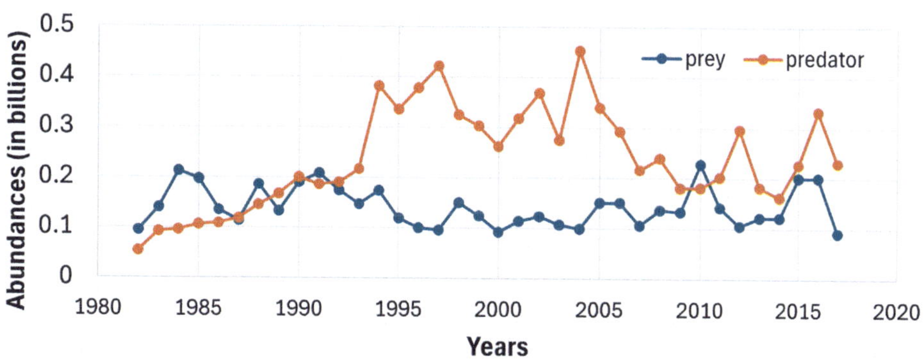

Fig. 3.7 Abundances of Atlantic menhaden and striped bass fish populations in millions (Data adapted from the Atlantic States Marine Fisheries Commission website [14])

earlier in Sect. 3.3.1—likely due to environmental variability and observation noise—we proceed, as before, to fit the generalized Lotka–Volterra model (3.6) to these field data. We need to point out that, because the fitting problem is inherently nonlinear and the mathematical system is overdetermined, fitting may lead to different parameter sets with comparably good fits to the time-series data. In particular this means that the solution to this problem is not unique.

We apply the same method discussed in the previous section using *R*, with the full program provided on GitHub [8] for student use. After creating the per-capita growth rates for each species, we run linear regression model `lm` for each of the two equations we are fitting. The regression analysis of the generalized Lotka–Volterra model revealed contrasting dynamics between prey (Atlantic menhaden) and predator (striped bass).

Summary for Prey Model

```
Call:
lm(formula = dNNdt_Prey ~ prey + pred, data = prey_lmData)

Coefficients:
            Estimate Std. Error t value Pr(>|t|)
(Intercept)   1.1135     0.2162   5.151 1.28e-05 ***
prey         -5.8681     1.1054  -5.309 8.09e-06 ***
pred         -1.1263     0.4222  -2.668   0.0119 *
---
Signif. codes:  0 '***' 0.001 '**' 0.01 '*' 0.05 '.' 0.1 ' ' 1

Residual standard error: 0.2402 on 32 degrees of freedom
```

(continued)

```
  (1 observation deleted due to missingness)
Multiple R-squared:  0.4796,Adjusted R-squared:  0.4471
F-statistic: 14.75 on 2 and 32 DF,  p-value: 2.889e-05
```

As seen in the results from the summary above, for menhaden, the intrinsic per-capita growth rate is positive and highly significant ($r = 1.11$, $p < 0.001$). Both self-limitation ($a = -5.87$, $p < 0.001$) and predation by striped bass ($b = -1.13$, $p = 0.012$) exerted significant negative effects, and the model explained approximately 48% of the variation in per-capita growth rates ($R^2 = 0.48$). In contrast, the striped bass regression indicated a positive baseline growth rate ($e = 0.57$, $p = 0.008$) and strong self-limitation ($c = -1.44$, $p < 0.001$), but the effect of prey abundance on predator growth is negative and not statistically significant ($d = -1.29$, $p = 0.22$). Overall, the striped bass model explained only about 30% of the variation in growth rates ($R^2 = 0.30$). These results suggest that menhaden dynamics are strongly shaped by both crowding and predation, whereas striped bass are primarily regulated by intra-specific density dependence, with little detectable response to menhaden abundance in this framework.

Summary for Predator Model

```
Call:
lm(formula = dNNdt_Pred ~ pred + prey, data = pred_lmData)

Coefficients:
            Estimate Std. Error t value Pr(>|t|)
(Intercept)   0.5728     0.2019   2.838 0.007824 **
pred         -1.4423     0.3942  -3.659 0.000904 ***
prey         -1.2935     1.0322  -1.253 0.219213
---
Signif. codes:  0 '***' 0.001 '**' 0.01 '*' 0.05 '.' 0.1 ' ' 1

Residual standard error: 0.2243 on 32 degrees of freedom
  (1 observation deleted due to missingness)
Multiple R-squared:  0.2954,Adjusted R-squared:  0.2514
F-statistic: 6.708 on 2 and 32 DF,  p-value: 0.00369
```

The results from these regressions show a classic asymmetric relationship: prey growth is significantly reduced by predators, but predator growth does not show a statistically significant positive response to prey. This mismatch is common in ecological field data, where predators are regulated by multiple factors beyond just the focal prey. These regression results are consistent with field observations indicating that striped bass are no longer relying only on Atlantic menhaden as a primary food source. Declines in

Atlantic menhaden abundance have been documented in recent decades, and striped bass have increasingly relying on alternative prey such as bay anchovy, river herring, and white perch [15]. This dietary diversification explains why the effect of menhaden abundance on striped bass growth (d) was not statistically significant, while strong intra-specific limitation (c) remained evident. Thus, the two-species generalized Lotka-Volterra framework captures the negative feedback within each population but underestimates the broader food-web complexity shaping predator dynamics in the Chesapeake Bay. This means that a multi-species model including additional prey species is needed to better capture predator's dynamics.

To evaluate model performance, we visualize the simulated trajectories against the observed abundances, as shown in Fig. 3.8. The fitted trajectories capture the broad trends in both series but smooth over much of the year-to-year variability. For Atlantic menhaden, the model (black line) reflects the decline and subsequent stabilization after the early 1980s, while the observed points fluctuate more widely around this level. For striped bass, the fitted curve (red line) shows a logistic-like rise and plateau consistent with the data's overall recovery, yet it underestimates the amplitude of inter-annual variations. Together, these patterns suggest that while the model is reasonable for projecting the long-term dynamics, it does not reproduce short-term predator-prey oscillations, likely due to environmental variability, harvest effects, and the limitations of a linear regression. It illustrates why more sophisticated approaches—like nonlinear optimization, sensitivity analysis, or stochastic modeling—are needed to align model predictions with observed variability.

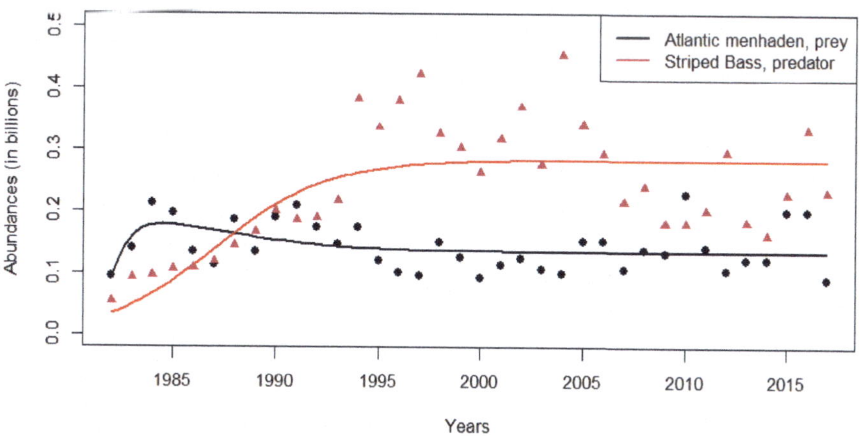

Fig. 3.8 Generalized Lotka–Volterra fit to Atlantic menhaden and striped bass time-series data in *R* **without optimization**. Atlantic menhaden: the model is shown in black solid line and data are represented black circles. Striped bass: the model is shown in red solid line and data are represented with red triangles

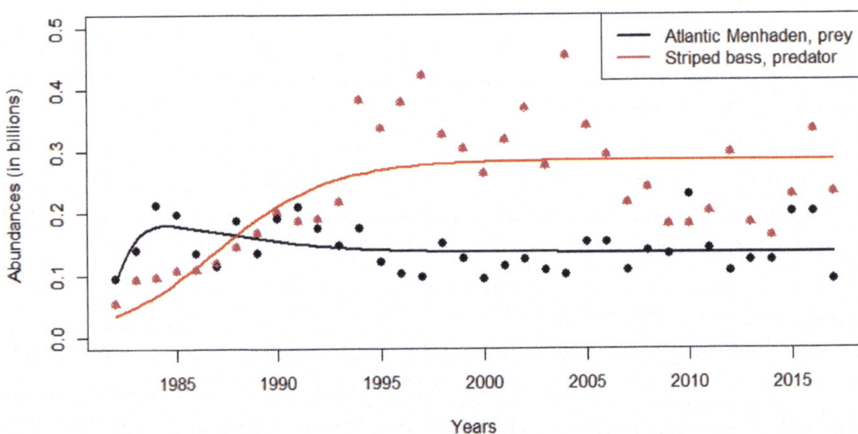

Fig. 3.9 Generalized Lotka–Volterra fit to Atlantic menhaden and striped bass time-series data in *R* **with optimization**. Atlantic menhaden: the model is shown in black solid line and data are represented black circles. Striped bass: the model is shown in red solid line and data are represented with red triangles

Table 3.3 Table of the model's parameter values obtained from data fitting using linear regression (no optimization) and with optimization. For each case the equilibrium values (x^*, y^*) are provided

Parameters	r	a	b	e	c	d	x^*	y^*
Value, no optim	1.1	−5.9	−1.10	0.6	−1.4	−1.3	0.1348	0.2833
Value, with optim	1.9	−9.5	−2.0	0.5	−1.5	−0.5	0.1351	0.2847

To this end, we run the second part of the code, which applies the nonlinear optimization routine using the `optim` command in *R*. Unlike the linear regression approach, this method directly minimizes the difference between model trajectories and the observed time series. The fitted curves obtained from this procedure are shown in Fig. 3.9. When compared to the regression-based fit, the results are very similar: both methods capture the broad stabilization of Atlantic menhaden abundances to about 135 millions and the rise and plateau of striped bass populations, which stabilizes at about 283 millions. The close agreement reflects the robustness of the fitted trends, but also highlights that, in this case, the added complexity of nonlinear optimization produces only modest visible improvement in the overall curves. However, the optimized method is more reliable, as it directly minimizes the mismatch between model dynamics and data, rather than relying on linear approximations. This provides a better foundation for subsequent analyses, such as parameter sensitivity.

Table 3.3 summarizes the parameter estimates obtained from both the linear regression and nonlinear optimization methods. The values are reported with their respective signs to indicate the direction of each interaction. The table also includes the corresponding equilibrium values associated with each fitted system.

The parameter values tell us how strongly each species influences itself and the other species—whether through self-limitation, predation, or competitive effects. The sign of a parameter (positive or negative) determines whether that effect promotes or suppresses growth. The equilibrium values (x^*, y^*), on the other hand, represent the long-term population levels at which growth and decline balance out. Biologically, these equilibria can be interpreted as the *steady states* of the system, where species abundances no longer change unless disturbed. In terms of abundance, the steady state of Atlantic menhaden is at about 135 million and the striped bass at about 284 million.

Interestingly, the equilibrium values obtained from the linear regression and nonlinear optimization methods are very close to each other. This suggests that, despite differences in parameter estimation techniques, both methods converge to a similar picture of the system's long-term behavior. In other words, the equilibria are robust features of the model, lending confidence to their biological interpretation.

Takeaway for Students Both approaches give similar-looking results here, but they serve different purposes. Linear regression is fast and useful for exploration, while nonlinear optimization provides a more biologically interpretable and mathematically consistent fit. In practice, nonlinear optimization is the method of choice when studying predator–prey dynamics, especially if we want to use the obtained model to make predictions or explore management strategies.

3.3.3 Resilience of a Predator-Prey System

Resilience is a measure of local stability: it quantifies the characteristic time a system needs to return to its equilibrium after a small perturbation. Following Pimm and Lawton [16], if λ_{max} denotes the leading eigenvalue of the Jacobian at the equilibrium (the one with largest real part), then for a locally asymptotically stable equilibrium we have $Re(\lambda_{max}) < 0$ and the return time is given by

$$\tau = -\frac{1}{Re(\lambda_{\max})}. \tag{3.8}$$

Thus, the resilience, defined as $\mathcal{R} = \dfrac{1}{\tau}$, is well-defined (and positive) only when the equilibrium is stable. As the $\mathcal{R}$ and τ are inversely proportional, the higher the resilience of the system, the lower the time for its return to equilibrium when perturbed from its steady state.

Let us use the obtained values from Table 3.3 in the case of linear regression model to find the eigenvalues of the Jacobian matrix evaluated at the equilibrium point of the generalized Lotka–Volterra predator–prey system. By substituting the equilibrium ($x^* = 0.1348$, $y^* = 0.2833$) and the corresponding parameter values into the Jacobian, we compute the eigenvalues $\lambda_1 = -0.9039$, $\lambda_2 = -0.2237$. Both are real and negative, indicating a locally asymptotically stable node at that point. From these eigenvalues, we can calculate characteristic decay time $\tau = 4.5$, meaning perturbations decay on a time scale of about 4.5 units (years). All calculations are carried out in MATLAB, and the full program used for this analysis is provided in the project's GitHub [8] repository for reference and reproducibility.

3.4 Parameter Sensitivity Analysis

Sensitivity analysis is an essential step in mathematical model development, providing insight into how small variations in the input parameters affect the model outputs [17, 18]. It helps identify parameters with minimal influence for model reduction, as well as those strongly correlated with the outputs, revealing directional effects and key drivers of system behavior. Sensitivity analysis also highlights parameters contributing most to output uncertainty, which may be asymmetric—such as when lower bounds can be estimated with confidence while upper bounds remain uncertain. By clarifying the consequences of parameter changes, it supports informed decision-making, robust scenario evaluation, and improved model reliability. As emphasized by [19], sensitivity analysis is now considered a standard requirement for good modeling practice in systems analysis and policy support. An extended version of this section by the same authors appears in [20].

Important: Sensitivity Analysis as a Teaching Tool

In this section we present a hands-on introduction on parameter sensitivity in mathematical models built with ordinary differential equations. Our focus is on making the concepts approachable, and accessible for student exploration. Beyond its central importance in mathematical modeling, sensitivity analysis can be an effective teaching tool. It invites students to engage directly with models—experimenting with parameters, testing hypotheses, and discovering how changes influence system behavior. We introduce the essential ideas behind both local and global sensitivity approaches, including several widely used techniques. Through this process, students strengthen their problem-solving abilities and learn to pinpoint the variables that matter most, enabling them to make decisions grounded in evidence. Such activities offer a practical setting in which to apply concepts from calculus, differential equations, and statistics, while also highlighting the interdisciplinary nature of mathematical applications. In doing so, sensitivity analysis not only deepens understanding of mathematics but also equips students with transferable skills to address complex, real-world challenges in their future careers.

3.4.1 Local Sensitivity Analysis

We first start with the **direct differential method**, a local approach to parameter sensitivity analysis that quantifies how infinitesimal changes in a parameter affect the model's output near a specified parameter value. Classified as "local" because it focuses on behavior in the immediate neighborhood of a given point in parameter space, this method computes partial derivatives of the model output with respect to each parameter, often through numerical differentiation. This approach provides a precise, point-specific measure of sensitivity, offering insight into which parameters exert the greatest influence on the system under baseline conditions.

Let us consider the model describing the population dynamics of the prey, Atlantic menhaden, $x = x(t)$ and the predator, striped bass, $y = y(t)$, with included harvesting of both, the prey and the predator:

$$\begin{cases} \dfrac{dx}{dt} = x(r - ax - by) - q_1 E_1 x \\ \dfrac{dy}{dt} = y(-e - cy + dx) - q_2 E_2 y. \end{cases} \tag{3.9}$$

Here, r is the intrinsic growth rate of the prey, e is the death rate of the predator in absence of prey, a and c are self-limitation parameters for the prey and the predator respectfully, b is the predator effect of y on x, and d is the effect of prey consumption of x on y. It is assumed that both the prey and the predator are harvested based on the catch per effort hypothesis. Parameters E_1 and E_2 are the harvesting efforts of the Atlantic menhaden and stripped bass respectively. Parameters q_1 and q_2 represent the catchability coefficients of each species. The range of catchability parameters is $[0, 1]$ with lower values used when the fish is relatively easy to catch, and higher values when the fish is more rare or difficult to catch. For simplicity, we assume $q_1 = 0.3$ and $q_2 = 0.9$ to be constants and ignore their variability given our limited control over their values. In contrast, parameters E_1 and E_2 offer us the flexibility to manipulate and control harvesting. In computations we chose $E_1 = 0.1$ and $E_2 = 0.1$.

For the remaining parameters (r, a, b, e, c, d), we use the values reported by Panayotova et. al in [21], who fitted the Lotka–Volterra model to the 1995–2017 subset of the same dataset we introduced in Sect. 3.3.2 employing more sophisticated optimization methods in Python to capture inter-annual (year-to-year) variability. We adopt their estimates because they better reflect observed predator-prey dynamics and, importantly, yield an intrinsic growth rate for Atlantic menhaden of $r = 0.513$, which is of similar magnitude to the growth rate estimated by [22] from observational data of Atlantic menhaden. These parameters, summarized in Table 3.4, will serve as the baseline values for the sensitivity analysis in this section.

Choosing the initial conditions $x(0) = 0.16$ and $y(0) = 0.31$ billions, corresponding to the abundances in year 2000, we get time-dependence of quantities x and y presented in Fig. 3.10.

Table 3.4 Table of the model's parameter values obtained from data fitting by Panayotova et al. in [21]

Parameter	r	a	b	e	c	d
Value	0.513	0.026	1.765	1.213	0.520	9.999

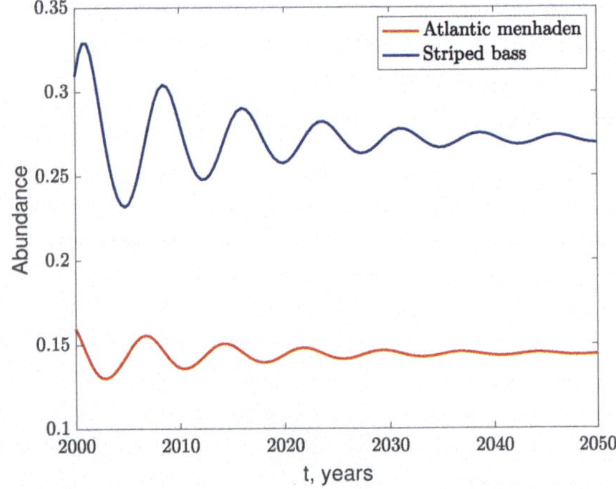

Fig. 3.10 Model prediction of the population dynamics of Atlantic menhaden and striped bass up to year 2025 using parameters from Table 3.4. Abundances are given in billions

Let us rewrite equation (3.9) in a vector form

$$\frac{d\boldsymbol{Z}}{dt} = \boldsymbol{F}(\boldsymbol{Z};\, \boldsymbol{p}), \quad \boldsymbol{p} = [r;\, a;\, b;\, e;\, c;\, d;\, E_1;\, E_2], \tag{3.10}$$

where $\boldsymbol{Z} = [x(t);\, y(t)]$ is the two-component vector solution of the system (3.9), and all parameters are represented as components of a vector $\boldsymbol{p}$. Consequently, our model is defined by a total of eight parameters: $\boldsymbol{p} = [r;\, a;\, b;\, e;\, c;\, d;\, E_1;\, E_2]$. We define sensitivity vector $\boldsymbol{s}_j$ for parameter p_j as $\boldsymbol{s}_j = [s_{1,j};\, s_{2,j}] = \left[\frac{\partial x}{\partial p_j};\, \frac{\partial y}{\partial p_j}\right]$, where $j = 1, 2, ..., 8$. Each of its components describes how a variation in p_j affects either x or y. These sensitivity vectors form the matrix of sensitivity measures

$$\boldsymbol{S} = \begin{bmatrix} \frac{\partial x}{\partial r} & \frac{\partial x}{\partial a} & \frac{\partial x}{\partial b} & \frac{\partial x}{\partial e} & \frac{\partial x}{\partial c} & \frac{\partial x}{\partial d} & \frac{\partial x}{\partial E_1} & \frac{\partial x}{\partial E_2} \\ \\ \frac{\partial y}{\partial r} & \frac{\partial y}{\partial a} & \frac{\partial y}{\partial b} & \frac{\partial y}{\partial e} & \frac{\partial y}{\partial c} & \frac{\partial y}{\partial d} & \frac{\partial y}{\partial E_1} & \frac{\partial y}{\partial E_2} \end{bmatrix}. \tag{3.11}$$

If the exact solutions for x and y are known, we could find each of the components of the matrix $\boldsymbol{S}$ explicitly, and it would answer our questions regarding parameter sensitivity. Since it is not the case, we consider the time evolution of each of the sensitivity measures given as follows

$$\frac{ds_{i,j}}{dt} = \frac{d}{dt}\left(\frac{\partial Z_i}{\partial p_j}\right) = \frac{\partial}{\partial p_j}\left(\frac{dZ_i}{dt}\right) = \frac{\partial F_i}{\partial x} \cdot \frac{\partial x}{\partial p_j} + \frac{\partial F_i}{\partial y} \cdot \frac{\partial y}{\partial p_j} + \frac{\partial F_i}{\partial p_j};$$

$$i = 1, 2;\ \ j = 1, 2, ..., 8.$$

Written in matrix form, the equation above will look like

$$\frac{d\mathbf{S}}{dt} = \mathbf{J}\mathbf{S} + \mathbf{f}, \tag{3.12}$$

where $\mathbf{J}$ is the Jacobian defined by the formula

$$\mathbf{J} = \begin{bmatrix} \frac{\partial F_1}{\partial x} & \frac{\partial F_1}{\partial y} \\ \frac{\partial F_2}{\partial x} & \frac{\partial F_2}{\partial y} \end{bmatrix} = \begin{bmatrix} r - 2ax - by - q_1 E_1 & -bx \\ dy & -e - 2cy + dx - q_2 E_2 \end{bmatrix}, \tag{3.13}$$

and $\mathbf{f}$ is the matrix formed by taking the derivatives of the right-hand sides of the equations in the system (3.9) with respect to the parameters:

$$\mathbf{f} = \begin{bmatrix} x & -x^2 & -xy & 0 & 0 & 0 & -q_1 x & 0 \\ 0 & 0 & 0 & -y & -y^2 & xy & 0 & -q_2 y \end{bmatrix}. \tag{3.14}$$

System of differential equations (3.12) defines how the sensitivity measures change in time.

Now we can formulate the initial value problem consisting of differential equations (3.10) and (3.12) together with their initial conditions

$$\begin{cases} \frac{d\mathbf{Z}}{dt} = \mathbf{F}(\mathbf{Z}; \mathbf{p}), \\ \frac{d\mathbf{S}}{dt} = \mathbf{J}\mathbf{S} + \mathbf{f}, \\ \mathbf{Z}(0) = \mathbf{Z_0}, \mathbf{S}(0) = \mathbf{S_0}. \end{cases} \tag{3.15}$$

It worth mentioning here that if no parameters appear in the initial condition of the state equations, then $\mathbf{S_0} = \mathbf{0}$ [17]. Now we have a direct method for calculating sensitivity measures. For example, the initial value problem that investigates the rate of change of x and y with respect to parameter b has the following form

$$\begin{cases} \frac{dx}{dt} = x(r - ax - by) - q_1 E_1 x \\ \frac{dy}{dt} = y(-e - cy + dx) - q_2 E_2 y \\ \frac{ds_{13}}{dt} = (r - 2ax - by - q_1 E_1) s_{13} - bx s_{23} - xy \\ \frac{ds_{23}}{dt} = dy s_{13} - (e + 2cy - dx + q_2 E_2) s_{23} \\ x(0) = x_0, \quad y(0) = y_0 \\ s_{13}(0) = s_{23}(0) = 0. \end{cases} \tag{3.16}$$

Notice that the first two differential equations are our original differential equations (3.9) while the next two equations come from (3.12) using the Jacobian and the third column of the matrix f since b is the third of our eight parameters. Similar to (3.16) we can formulate initial value problems defining the rate of change of dependent variables with respect to any other parameter. To account for the different units of measurement and varying magnitudes of parameters, we introduce the dimensionless relative sensitivity S_{ij}, which is defined as $S_{ij} = \frac{\partial Z_i}{\partial p_j}\frac{p_j}{Z_i}$. For instance, we have $S_{13} = \frac{\partial x}{\partial b}\frac{b}{x}$ and $S_{23} = \frac{\partial y}{\partial b}\frac{b}{y}$. To capture the time-dependent nature of sensitivity measures, we define the sensitivity index $||S_{ij}||_2$ as the magnitude of the corresponding norm for each relative sensitivity:

$$||S_{ij}||_2 = \sqrt{\sum_{k=1}^{N}\left(\frac{\partial x_i}{\partial p_j}(t_k)\frac{p_j}{x_i(t_k)}\right)^2}. \tag{3.17}$$

For example, $||S_{13}||_2 = \left\|\frac{\partial x}{\partial b}\frac{b}{x}\right\|_2 = \sqrt{\sum_{k=1}^{N}\left(\frac{\partial x}{\partial b}(t_k)\frac{b}{x(t_k)}\right)^2}$, etc.

Computed values of relative sensitivities $S_{ij} = \frac{\partial Z_i}{\partial p_j}\frac{p_j}{Z_i}$ $(i = 1, 2; j = 1, ..., 8)$ and their associated sensitivity indices $||S_{ij}||_2$ are presented in Table 3.5. Computations were performed over a time period spanning from 2000 to 2080, with a duration of $t_f = 80$ years.

From the sensitivity indices summarized in in Table 3.5 we can conclude that the predator (striped bass) experiences the greatest sensitivity to the prey's intrinsic growth rate r and to the predation coefficient b. In contrast, the prey (Atlantic menhaden) is more sensitive to the predator's self-regulation term d and the predator's baseline mortality e, reflecting how changes that reduce predator pressure feed back to prey dynamics.

Table 3.5 Relative sensitivities & relative sensitivity indices

Parameter p	$\frac{\partial x}{\partial p}\frac{p}{x}$, ($t_f = 80$ years)	$\left\|\frac{\partial x}{\partial p}\frac{p}{x}\right\|_2$	$\frac{\partial y}{\partial p}\frac{p}{y}$, ($t_f = 80$ years)	$\left\|\frac{\partial y}{\partial p}\frac{p}{y}\right\|_2$
r	0.1176	3.0598	1.0647	18.1783
a	−0.0008	0.0202	−0.0077	0.1323
b	−0.1000	2.9838	−1.0048	16.2097
e	0.8502	15.4344	−0.0136	6.6165
c	0.0999	1.8899	−0.0014	1.1786
d	−1.0019	17.6650	0.0055	6.4200
E_1	−0.0069	0.1789	−0.0623	1.0618
E_2	0.0631	1.1466	−0.0010	0.4910

Important: Local vs Global Sensitivity Analysis
The **direct differential (local) method** provides a computationally efficient way to quantify these effects by evaluating derivatives of the model outputs with respect to parameters at a chosen baseline. This yields quantitative insight into how small perturbations of r, a, b, e, c, d influence equilibria, stability (via eigenvalues), and recovery times τ. Its flexibility makes it easy to tailor to specific model structures and questions, and its locality supports intuitive interpretation.

At the same time, it is important to recognize the limitations of the direct differential method. It presumes a smooth, differentiable model—an assumption that holds for our system but may fail for more complex models. Moreover, as a local sensitivity approach, it examines behavior only in the neighborhood of a chosen parameter set and therefore may not capture responses arising under different parameter combinations. Such local analyses can miss interaction effects and other phenomena that emerge when multiple parameters vary jointly. To obtain a broader perspective, we next turn to **global sensitivity analysis**, which evaluates model responses to simultaneous variation of all parameters across their admissible ranges. By doing so, we can reveal nonlinear behaviors and intricate parameter dependencies that local methods may not capture.

3.4.2 Basic Ideas of Global Sensitivity Analysis

We will present several **global sensitivity analysis (GSA) methods** and illustrate their use through the case study of the generalized Lotka-Volterra model (3.9) using parameters from Table 3.4. We start by outlining basic concepts characteristic of all methods.

Let the quantity of interest (QoI) be denoted by z for a model with m parameters $p_1, p_2, \ldots, p_m$. In our case study, the QoI is either the prey abundance x or the predator abundance y. From Eq. (3.10), the number of parameters is $m = 8$.

Global sensitivity analysis involves several typical steps:

1. **Specify parameter ranges and distributions**. Begin by setting plausible ranges informed by prior knowledge or data, and assign probability distributions to each parameter. A uniform distribution is common when prior information is limited, whereas a normal distribution can capture bell-shaped variability, symmetric behavior, or known constraints on values.
2. **Construct a sample of parameter sets**. Generate a collection of parameter vectors. For large or nonlinear models, classical analytic tools (e.g., linear stability analysis) may be impractical for capturing interaction effects. While local approximations can illuminate single-parameter influences, assessing joint effects typically requires a probabilistic

design. Random sampling avoids biases from systematic sweeps and promotes even exploration of the space. The sample size depends on factors like model complexity, desired accuracy, and computational resources, aiming to balance between model intricacy and computational efficiency.

Two commonly used designs are Monte Carlo and Latin hypercube sampling (LHS) [17]. Monte Carlo draws each parameter p_i independently from its distribution, repeating to build many parameter sets. LHS, introduced by McKay et al. [23], partitions the range $[p_i^{\min}, p_i^{\max}]$ of each p_i into n equal-probability intervals and selects one random value from each; the n values for p_1 are randomly permuted and paired with those for p_2, then combined with the n values for p_3, and so on, yielding an $(n \times m)$ matrix P with n sampled rows and m parameter columns.

The advantage of LHS is its efficiency. Unlike standard Monte Carlo sampling, which may lead to over-sampling in some regions and under-sampling in others, LHS ensures more even and efficient exploration of the parameter space. This approach results in a better understanding of the model's behavior with fewer samples, saving time and computational resources. In MATLAB, functions `lhsdesign` (uniform) and `lhsnorm` (normal) support LHS sampling.

3. **Simulate the model for each parameter set to obtain outputs**. For each row of the matrix P, run the model to produce the corresponding output, yielding n simulated solutions.
4. **Visualize and explore model outputs.** Use plots (e.g., scatter plots, pairwise correlation maps) to examine simulation outputs across sampled parameter sets. Such diagnostic visuals help identify dominant trends, nonlinearities, and potential parameter interactions, thereby guiding the choice of appropriate GSA methods.
5. **Apply global sensitivity analysis method**. Use the simulated input–output data to quantify how variability in parameters affects the QoI. Method choice depends on variable types, assumed relationships, and data availability.
6. **Interpret results.** Identify influential parameters and characterize their effects on the model outputs.

To illustrate how different approaches capture distinct aspects of parameter influence, we next describe several widely used GSA methods: Pearson and Spearman correlations, partial rank correlation coefficients (PRCC), and Sobol' indices. Before proceeding with these methods, we briefly recall a few important statistical definitions.

Definition 3.2 The **variance** of a random variable measures how far values typically deviate from the mean. For a sample $p_1, \ldots, p_n$ with sample mean $\bar{p}$, a common "biased" estimator is

$$Var_b(p) = \frac{1}{n}\sum_{j=1}^{n}(p_j - \bar{p})^2,$$

where n is the sample size. In practice, it is preferable to use the **unbiased** sample variance,

$$Var(p) = \frac{1}{n-1}\sum_{j=1}^{n}(p_j - \bar{p})^2,$$

which applies Bessel's correction. Note that $Var_b(p)$ systematically underestimates the population variance; hence the unbiased form is typically used (see [24, 25] for further details).

Definition 3.3 The **standard deviation** of a random variable p, denoted σ_p, is the nonnegative square root of its variance:

$$\sigma_p = \sqrt{\mathrm{Var}(p)}.$$

Definition 3.4 The **covariance** between two random variables p and z quantifies the extent to which they co-vary. For a sample $(p_1, z_1), \ldots, (p_n, z_n)$ with means $\bar{p}$ and $\bar{z}$, a common "biased" estimator is

$$\mathrm{Cov}_b(p, z) = \frac{1}{n}\sum_{j=1}^{n}(p_j - \bar{p})(z_j - \bar{z}).$$

When analyzing data, it is standard to use the **unbiased** sample covariance,

$$\mathrm{Cov}(p, z) = \frac{1}{n-1}\sum_{j=1}^{n}(p_j - \bar{p})(z_j - \bar{z}),$$

which corrects the downward bias that arises from estimating the means from the same data (Bessel's correction); hence the $n - 1$ denominator is preferred.

3.4.3 Pearson Correlation Coefficients

In the context of global sensitivity analysis, Pearson's coefficients (Pearson correlation coefficients) assess the *strength* and *direction* of the linear relationship between model inputs and outputs [17].

Definition 3.5 Mathematically, **Pearson's coefficient** ρ is defined as the covariance between two variables p and z divided by the product of their standard deviations:

$$\rho = \frac{\mathrm{Cov}(p, z)}{\sigma_p \sigma_z}.$$

The coefficient range is $\rho \in [-1, 1]$: $\rho = -1$ indicates a perfect negative linear relationship, $\rho = +1$ a perfect positive linear relationship, and $\rho = 0$ no linear relationship. Positive values indicate that larger values of one variable tend to occur with larger values of the other, while negative values indicate the opposite.

In our case study, we are interested in the correlation between the quantity of interest z and each of the m parameters $p_1, p_2, ..., p_m$. So, we define the Pearson correlation coefficients as

$$\rho_{p_{i,z}} = \frac{\sum_{j=1}^{n}(p_{ij} - \bar{p}_i)(z_j - \bar{z})}{\sqrt{\sum_{j=1}^{n}(p_{ij} - \bar{p}_i)^2 \sum_{j=1}^{n}(z_j - \bar{z})^2}}, \quad i = 1, 2, ..., m \tag{3.18}$$

In formula (3.18) $\bar{p}_i$ and $\bar{z}$ represent the mean of p_i and z respectively. The value of $\rho_{p_{i,z}}$ can change between -1 and 1.

Below, we compute and compare Pearson correlation coefficients for each parameter in system (3.9) to rank their importance and understand their influence on the system's behavior. Various software packages offer the functionality to compute Pearson correlation coefficients. For instance, MATLAB provides a function called *corrcoef* that can be used for this purpose.

The p-value associated with each correlation coefficient indicates whether the observed correlation is statistically significant. A low p-value, typically below 0.05, indicates a strong correlation and provides evidence to reject the null hypothesis of no correlation between the variables. For a more in-depth understanding of hypothesis testing and p-values, interested readers are encouraged to explore relevant statistical textbooks, such as [24] or [25].

As mentioned earlier, one should start with generating a sample space of parameter sets. For our analysis, we employ LHS sampling, which ensures that each parameter $p_i (i = 1, 2, ..., 8)$ is sampled within an interval of $\pm 5\%$ of its nominal value provided in Table 3.4. This interval is defined as $[0.95 p_i, 1.05 p_i]$. Since we lack information about the probability distribution of the parameters, we opt for uniform sampling. We conducted sensitivity analyses with sample sizes of $n = 50$ and $n = 150$ for this simple model. The results did not show noticeable variation across these different sample sizes.

Let us assume that the quantity of interest (QoI) is the predicted population of the Atlantic menhaden x in 2050 ($t_f = 50$ years) and compare the Pearson correlation coefficients for parameters $r, a, b, e, c, d, E_1, E_2$. Notice that catchabilities q_1 and q_2 are assumed to be fixed.

Analyzing the results presented in Fig. 3.11, we find that the most crucial parameters for the Atlantic menhaden are e and d. These findings are supported by small p-values associated with correlation coefficients which indicate statistical significance. Specifically, the abundance of the prey (Atlantic menhaden) and the death rate e of the predator exhibit a positive correlation. Additionally, there exists a negative correlation between the abundance of the prey and parameter d, which is responsible for the effect of the prey

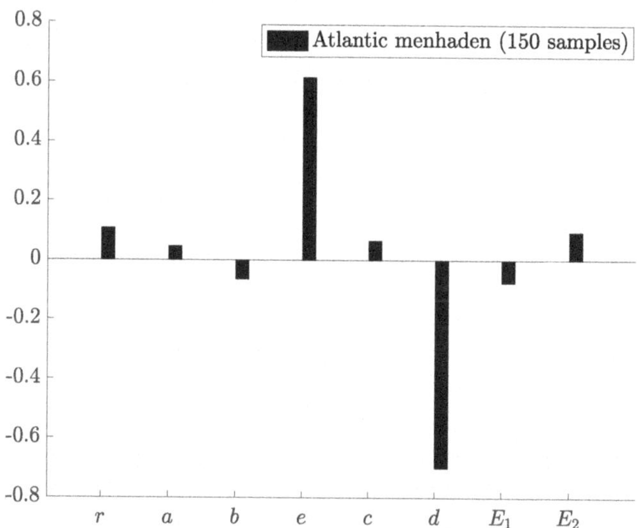

Fig. 3.11 Pearson's sensitivity coefficients for the Atlantic menhaden (n = 150 samples)

consumption on the predator. The effect of variations of all other parameters does not seem to be very noticeable.

Do the results shown in Fig. 3.11 imply that the other parameters are less significant? It is likely, but we should approach this with caution and consider the limitations of the Pearson correlation coefficient method.

Although Pearson's coefficients provide useful insight, they quantify only linear relationships and can misrepresent nonlinear input–output dependencies in global sensitivity analysis. To address this limitation we create scatter-plots of each parameter against the QoI as they offer valuable complementary information.

Definition 3.6 A **scatter-plot** is a graphical representation that allows us to visually observe patterns, trends, and potential non-linear associations between two variables by plotting their data points in a coordinate system. The horizontal axis represents the values of the input parameter $p_i (i = 1, 2, ..., m = 8)$, and the vertical axis represents the corresponding values of the QoI.

Figure 3.12 shows scatter-plots for parameters e, d, r, and E_2. Visual inspection indicates a noticeable positive correlation between the QoI and the parameter e (see panel (a)) and negative correlation between QoI and parameter d (see panel (b)). By contrast, the remaining parameters in the figure r, and E_2 display substantial variability with no clear correlation pattern. The scatter-plots for parameters a, b, c, and E_1 are omitted, as they likewise show wide dispersion without an apparent trend.

Now let us change the quantity of interest: we take the predicted population of striped bass y in 2050. We again compute and compare Pearson correlation–based sensitivity

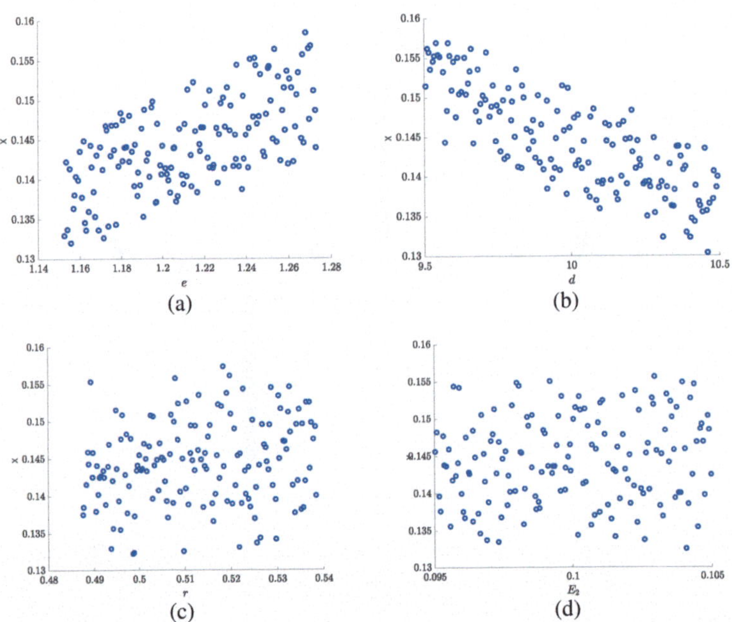

Fig. 3.12 Scatter-plots for: (**a**) parameter e, (**b**) parameter d, (**c**) parameter r, and (**d**) parameter E_2 against the value of x in 2050

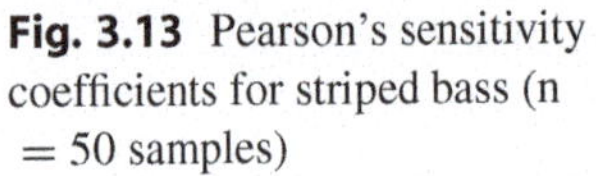

Fig. 3.13 Pearson's sensitivity coefficients for striped bass (n = 50 samples)

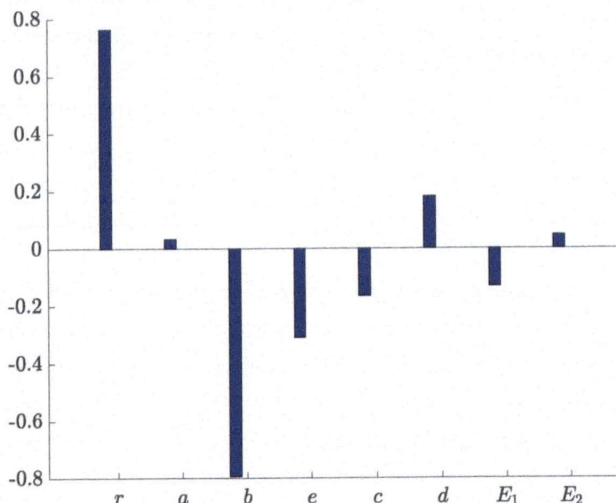

coefficients for parameters $r, a, b, e, c, d, E_1, E_2$, keeping the catchabilities q_1 and q_2 fixed.

From Fig. 3.13, the most influential parameters for striped bass are r and b. This conclusion is supported by the p-values associated with the correlation coefficients, which indicate statistical significance (p-value < 0.05). The abundance of the predator (striped bass) is positively correlated with the prey growth rate r, and negatively correlated with

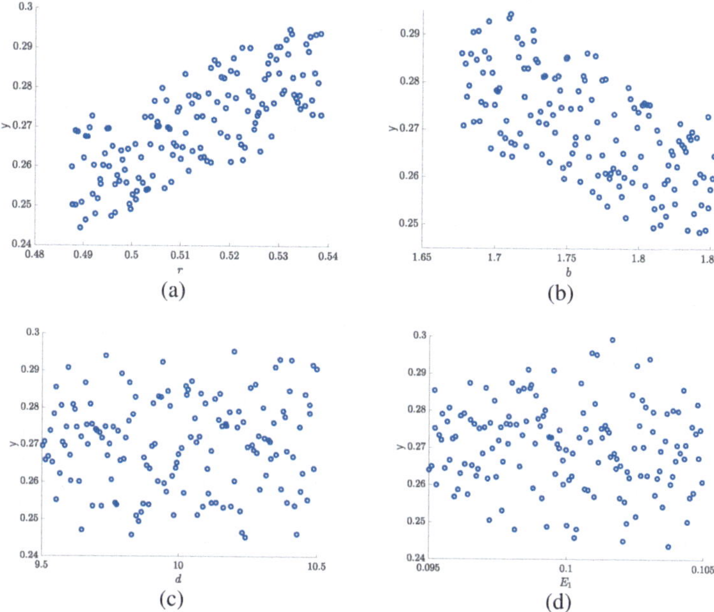

Fig. 3.14 Scatter-plots for: (**a**) parameter r, (**b**) parameter b, (**c**) parameter d, and (**d**) parameter E_1 against the value of y in 2050

the parameter b governing the predator's effect on the prey. The effects of the remaining parameters appear less pronounced; nevertheless, we examine the corresponding scatter-plots before drawing firm conclusions.

Figure 3.14 displays scatter-plots for parameters r, b, d, and E_1 against the QoI which is now the value of y at $t_f = 50$ years. It reveals a clear correlation between the QoI and parameters r and b. At the same time, the remaining parameters exhibit significant variations with no clear correlation pattern. The scatter-plots for parameters a, c, e, and E_2 are not included, as they demonstrate substantial variations without noticeable pattern.

Important: Limitations of the Pearson's Correlation

Pearson's correlation is most informative for *linear* relationships. When the relationship between a parameter and the QoI is nonlinear, its interpretation becomes problematic. In such settings, a fitted slope is not meaningful and ρ can be misleading. In the case when the dependence between parameters and model outcomes is monotonic, a practical workaround is to rank-transform the variables, substituting each value by its ordinal rank. This emphasizes monotonic structure even when the form is nonlinear. The following section introduces **Spearman correlation**, an alternative based on Pearson's ρ applied to the ranked data.

3.4.4 Spearman Correlation

Because Pearson's correlation reflects linear association, nonlinear parameter–QoI relationships can make it hard to interpret. A useful alternative is to apply a rank transform, replacing each value by its ordinal rank; this approach is appropriate when the dependence is monotonic.

Rank transformation is a technique used to convert the actual values of a dataset into their corresponding ranks. For a given dataset, first order the values from smallest to largest and then assign ranks $1, 2, ..., n$. If a set has equal values instead of consecutive rank positions, assign to each the average of those positions (the midrank). For example, for the set $\{10, 5, 8, 8, 3\}$ the ordered values are $(3, 5, 8, 8, 10)$, so the ranks are

$$3 \mapsto 1, \quad 5 \mapsto 2, \quad 8 \mapsto \tfrac{3+4}{2} = 3.5, \quad 8 \mapsto 3.5, \quad 10 \mapsto 5.$$

For monotonic input–output relationships, rank transformations of the input and output values (i.e. replacing the values with their ranks) results in linear relationships and the rank coefficient indicates the degree of monotonicity between the input and output values. The Spearman correlation coefficient therefore quantifies the strength and direction of the monotonic association between parameter values and the corresponding model outputs (QoI). Unlike Pearson's correlation, which assumes linearity, the Spearman correlation is based on the ranks of the data rather than the actual values. It assesses the extent to which the variables tend to change together in a consistent, monotonic manner, regardless of the specific functional form of the relationship. Positive Spearman correlation indicates that higher parameter values tend to be associated with higher QoI values, while negative Spearman correlation indicates an inverse relationship.

Spearman's rank correlation coefficient ρ_s is computed analogously to (3.18), except that it operates on rank-transformed data:

$$\rho_s = \operatorname{Corr}\big(R(p),\ R(z)\big),$$

where $R(\cdot)$ denotes (mid)ranks. Many software packages implement this calculation—for example, in MATLAB one can use `corr(...,'Type','Spearman')`. As with Pearson's correlation, it is common to report a p-value for testing $H_0 : \rho_s = 0$. **Low p-values** (usually below 0.05) signify a **strong correlation** and offers evidence to reject the null hypothesis of no correlation between the variables.

In the context of our case study, let's consider the task of determining the parameters that have the greatest impact on the quantity of interest (QoI), which is defined as the population of the striped bass x at time $t_f = 50$ years. We have a total of eight parameters to rank, resulting in an 8-dimensional sample space. To sample these parameters, we employ a LHS sampling centered at the nominal parameter values provided in Table 3.4, sampling each parameter $p_i (i = 1, 2, ..., 8)$ within the interval $[0.95p_i, 1.05p_i]$.

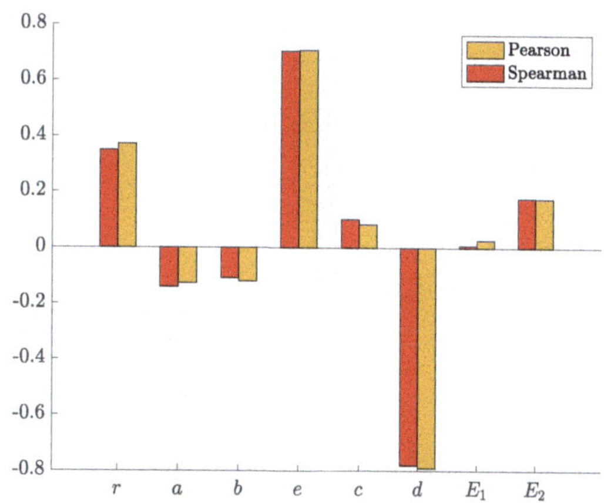

Fig. 3.15 Comparison of Pearson and Spearman correlations when the QoI is the value of x in 2050, assuming uniform distribution of the parameters

Because the parameter probability law is unknown, we adopt a uniform sampling scheme and analyze its outcomes. Calculations use a sample size of $n = 150$. Results for the uniform probability density function are shown in Fig. 3.15 when the QoI is the value of x. Observe that parameters e and d remain the most influential for prey abundance. Importantly, the signs of the correlation coefficients remain consistent, with negative correlations continuing to indicate an inverse relationship between the QoI and a parameter and positive correlations indicating a positive relationship, especially for the more influential parameters.

Exploring alternative distributional assumptions (e.g., normal sampling with a larger n) and comparing them with the uniform case is left for student practice. Although we present uniform-sampling results here, calculations with alternative distributions (see Problem 3.7) yield the same most sensitive parameters.

Figure 3.16 illustrates the comparison of Pearson and Spearman correlation coefficients, with the QoI being the population of striped bass y at time $t_f = 50$ years under the assumption of the uniform distribution of the parameters. The results demonstrate qualitative similarity between the two methods. As before, parameters r and b exhibit the greatest influence on predator abundance.

It is important to emphasize that monotonicity plays a significant role in interpreting correlation coefficients. For rank transformation to yield a linear-like relationship, the original data should exhibit monotonicity.

3.4.5 Partial Rank Correlation Coefficient (PRCC)

While Pearson's correlation captures linear (or near-linear) association between a parameter and the QoI, Spearman's correlation replaces values by ranks to handle nonlinear but *monotonic* relationships. However, both measures are *marginal*: they do not adjust for variability in the QoI induced by other parameters. To isolate the association between

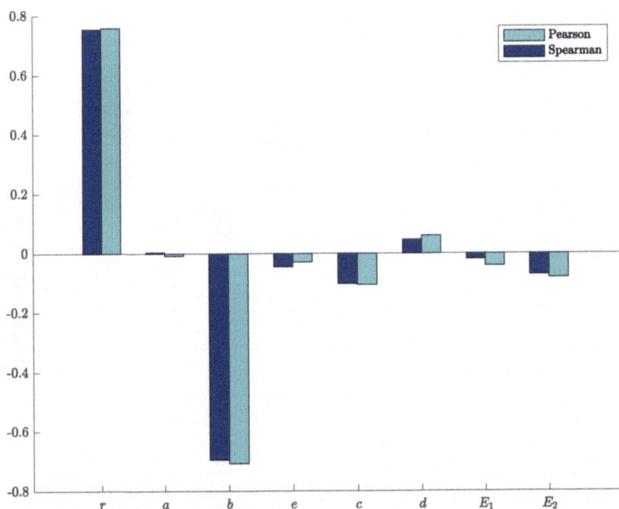

Fig. 3.16 Comparison of Pearson and Spearman correlations when the QoI is the value of y in 2050, assuming uniform distribution of the parameters

a single parameter and the QoI while accounting for the remaining inputs, we use *partial correlation coefficients* [26]. In this section we introduce the **Partial Rank Correlation Coefficient** (PRCC) as a robust sensitivity analysis method that combines ranked correlation and partial correlation.

Definition 3.7 Let $p_1, p_2, \ldots, p_n$ denote model inputs and let z be the quantity of interest (QoI). The partial rank correlation coefficient for p_1, denoted PRCC_{p_1}, measures the association between p_1 and z after adjusting for the remaining inputs. In the two-parameter case with inputs p_1 and p_2, PRCC_{p_1} captures the correlation between p_1 and z *conditional on* p_2, thereby discounting indirect associations mediated by relationships between p_1 and p_2 or between p_2 and z.

Mathematically the $PRCC_{p_1}$ is defined as:

$$PRCC_{p_1} = \frac{\rho_{p_1,z} - \rho_{p_1,p_2}\rho_{p_2,z}}{\sqrt{(1-\rho^2_{p_1,p_2})(1-\rho^2_{p_2,z})}} \tag{3.19}$$

In general, to discount all the interaction we subtract the sum of all covariances between p_1 and $p_j, j \neq 1$. Notice that the parameters in formula (3.19) were rank-transformed.

The partial rank correlation coefficient (PRCC) is widely used in sensitivity analysis because it is easy to implement and robust for isolating each parameter's contribution to model outcomes. A key advantage is that the *sign* of the coefficient retains an interpretable meaning even for nonlinear (but monotonic) relationships. As with Pearson and Spearman

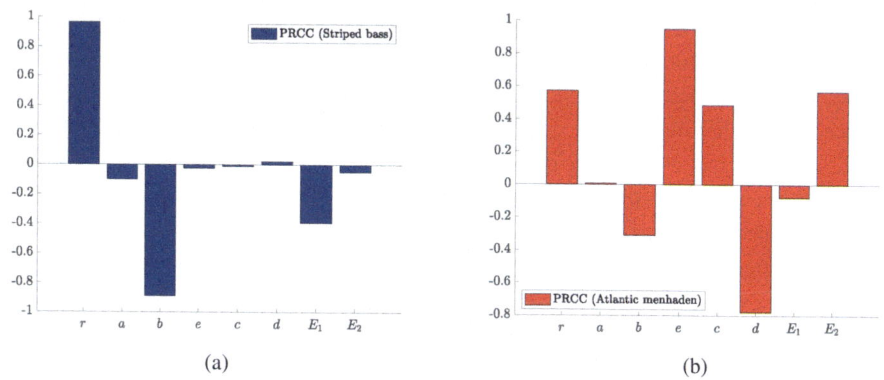

Fig. 3.17 PRCC values for ($t_f = 50$ years). (**a**) Striped bass. (**b**) The Atlantic menhaden

measures, PRCC values near +1 indicate a strong positive monotonic association between a parameter and the QoI (conditional on the other inputs), whereas values near −1 indicate a strong negative association. Statistical significance can be assessed via hypothesis tests of H_0 : PRCC = 0 with associated p-values; small p-values (<0.05) provide evidence against the null of no association. In MATLAB, partial correlations can be computed with the command `partialcorr`.

Figure 3.17 reports the PRCC results for our case study. We performed the analysis in an 8 -dimensional sample space using a LHS sampling with normal marginals centered at the nominal parameter values in Table 3.4, truncated to the interval $[0.95\ p_i,\ 1.05\ p_i]$ for each p_i $(i = 1, \ldots, 8)$. To select an adequate design size, we explored sample sizes n ranging from 10^2 to 10^3.

The PRCC values were computed for each of the eight input parameters with respect to the two outcome variables: (a) the predicted abundance of striped bass and (b) the predicted abundance of Atlantic menhaden, in 2050 ($t_f = 50$ years). Using the PRCC framework, we identified the parameters exerting the strongest influence on both species illustrated in Fig. 3.17. For striped bass as seen in panel (a), there is a robust positive correlation with r (beneficial effect on the QoI) and a clear negative correlation with b (adverse effect). Notably, the harvesting effort on menhaden E_1 had a p-value < 0.05 and a PRCC of approximately −0.4, indicating a moderate but statistically significant negative association with striped bass abundance.

When the QoI is the abundance of Atlantic menhaden, as we can observe in Fig. 3.17 panel (b), there is a strong positive correlation with e and a pronounced negative correlation with d. Parameters r, c, and E_2 show moderate yet statistically significant positive correlations with menhaden abundance. Comparing these PRCC results with the findings in Figs. 3.11, 3.13, 3.16, and 3.15 reveals strong qualitative agreement regarding which parameters have substantial, highly significant impacts on the QoI, reaffirming the consistency of our sensitivity analysis.

To examine sensitivity over time, we compute sensitivity indices at multiple time points and rank parameters for a range of final times t_f. This reveals how the system's evolution towards the steady state is influenced by the parameters. By performing sensitivity analyses at discrete times $(t_1, t_2, \ldots, t_k)$ and tracking partial rank correlation coefficients (PRCCs) across time, we identify dominant processes and their characteristic time scales. Rapid changes in early-times suggest that closely spaced observations may be needed to capture critical dynamics. Identifying the most sensitive parameters informs efficient control strategies, and time-varying rankings indicate when control policies should adapt over time.

Figures 3.18 and 3.19 show the temporal evolution of PRCC values for the abundances of striped bass and Atlantic menhaden, respectively (as QoI). The sensitivity rankings at early times differ from those observed over longer horizons. For striped bass, parameters d, e, and E_2 are initially comparable in importance to r and b, but their sensitivities decay markedly with time. Conversely, for Atlantic menhaden, r and b are influential at the outset, yet their sensitivity rankings diminish relative to e and d as the system evolves.

PRCC is a widely used straightforward ranking method that offers robustness and ease of implementation. However, it is important to note that the effectiveness of its correlation coefficients relies on the presence of monotonic relationships between parameters and the QoI, ensuring a linear relationship through rank transformation. This can require visual inspection of scatterplots and sometimes reduction of the parameter space. Also, it is a good practice to vary the number of samples in order to decide how many is sufficient.

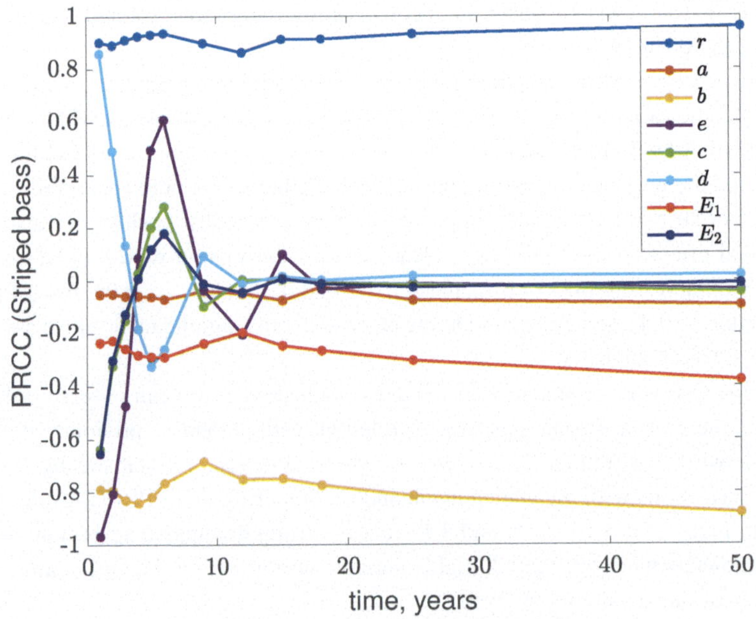

Fig. 3.18 PRCC values in time for the case when the QoI is the value of y

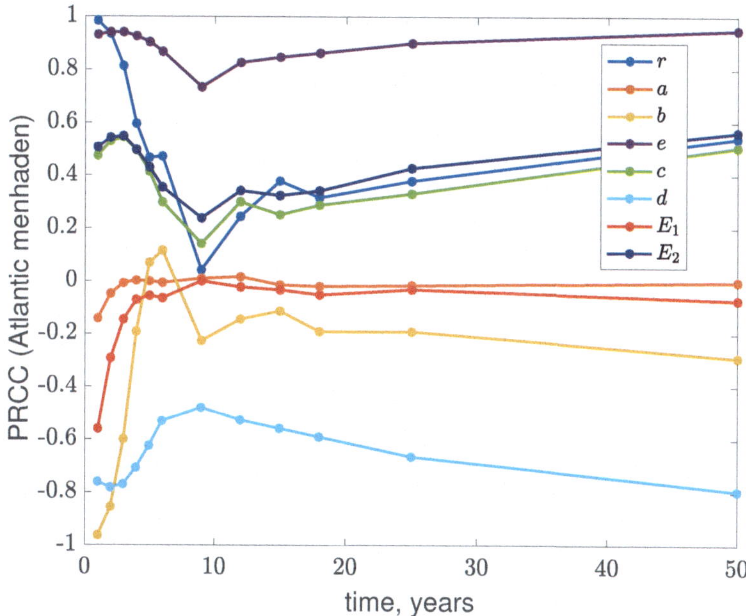

Fig. 3.19 PRCC values in time for the case when the QoI is the value of x

3.4.6 Method of Sobol'

We next discuss the sensitivity-analysis framework introduced by I. M. Sobol' (1993) [27]. The method is based on an ANOVA (Analysis of Variance)-style variance decomposition that partitions the output variance among the input parameters (and their interactions). Two features distinguish Sobol' analysis: (i) it applies naturally to nonlinear input–output relationships, and (ii) it quantifies parameter *interactions* in addition to main (first-order) effects. By contrast, PRCC—while often robust for monotonic relationships—is a marginal, rank-based measure that does not account for parameter interactions and focuses on the primary relationship between a single input and the QoI.

The Sobol' methodology differs fundamentally from the approaches above. Sobol' sensitivity is *variance-based*: it begins with the variance of the model output and partitions that variance among inputs and their interactions, rather than relating sensitivity to correlation. Correlation asks whether changes in a parameter co-vary with changes in the QoI; **variance-based sensitivity asks how much of the *output variance* is attributable to that parameter**. Thus, a parameter is influential in the Sobol' sense if **perturbations in it *induce large variations* of the QoI around its mean** [28].

ANOVA, originally introduced by Fisher [29], is a widely used statistical framework for assessing differences among group means. It partitions the total variability into components attributable to **variation *between* groups** and **variation *within* groups**, thereby **identifying sources of significant differences that affect the measurements under study**.

Sobol' [27, 30] generalized ANOVA to show that any suitable input—output general functional relationship $f(p_1, \ldots, p_m)$ can be expanded as a sum of orthogonal component functions. This *functional ANOVA* decomposition yields an associated partition of the total output variance into contributions from individual parameters and their interactions. Orthogonality ensures that the total variance splits additively into main-effect and interaction components, enabling a principled variance-based sensitivity analysis.

$$f(p_1, p_2, ..., p_m) = f_0 + \sum_{i=1}^{m} f_i(p_i) + \sum_{1 \le i < j \le m} f_{ij}(p_i, p_j) + ...f_{1,...,m}(p_1, ..., p_m). \tag{3.20}$$

If the input parameters are mutually independent, then there exist a unique decomposition (3.20) such that all the summands are mutually orthogonal i.e. if $(i_1, ..., i_s) \neq (j_1, ..., j_q)$, then

$$\int_{[0,1]^m} f_{i_1,...,i_s}(p_{i_1}, ..., p_{i_s}) f_{j_1,...,j_q}(p_{j_1}, ..., p_{j_q}) d\mathbf{p} = \mathbf{0} \tag{3.21}$$

From Eqs. (3.20) and (3.21) we can find a relationship between the total variance (Var) and variance due to parameters

$$Var = \sum_{i=1}^{m} V_i + \sum_{i<j} V_{ij} + \sum_{i<j<l} V_{i,j,l} + ... + V_{1,2,3,...,m}, \tag{3.22}$$

where $V_{i_1,i_2,...,i_s} = \int_{[0,1]^m} f^2_{i_1,i_2,...,i_s} dp_{i_1}...dp_{i_s}$ is the variance due to specific parameter combinations $(p_{i_1}, ..., p_{i_s})$.

It is more useful to compare the partial variances to the total variance dividing equation (3.22) by Var and rewriting it as

$$1 = \frac{\sum_{i=1}^{m} V_i}{Var} + \frac{\sum_{i<j} V_{ij}}{Var} + ... + \frac{\sum_{i<j<l} V_{i,j,l}}{Var} + \frac{V_{1,2,...,m}}{Var} \tag{3.23}$$

Definition 3.8 The **first-order Sobol' sensitivity index**, often denoted as S_i, quantifies the contribution of the individual input parameter p_i to the total variance of the model output regardless of its interactions with other parameters:

$$S_i = \frac{V_i}{Var}. \tag{3.24}$$

Definition 3.9 The **higher-order Sobol' sensitivity indices**

$$S_{i_1,...,i_s} = \frac{V_{i_1,i_2,...,i_s}}{Var} V, \; (s > 1)$$

extend the analysis beyond the individual input parameters to capture the interactions between multiple parameters. These indexes provide insights into the combined effects of parameter interactions on the output variance.

Definition 3.10 The **total Sobol' sensitivity index**, often denoted as S_{T_i}, represents the total contribution of parameter p_i to the output variance, considering both its individual effect and its interactions with other parameters. It quantifies the proportion of the total output variance that can be attributed to the variations in parameter p_i. **The total Sobol' index is defined as**

$$S_{T_i} = \frac{\sum_{i=1}^{m} V_i}{Var} + \frac{\sum_{i\neq j} V_{ij}}{Var} + \ldots + \frac{\sum_{i\neq j\neq l} V_{i,j,l}}{Var} + \frac{V_{1,2,\ldots,m}}{Var} \tag{3.25}$$

Below, we report the computed total Sobol' indices and use them to rank parameter sensitivities in our case-study model. Using a Monte Carlo design, we drew $n = 5000$ samples from the parameter space, evaluated the model at each draw, and then decomposed the output variance to obtain sensitivity indices that account for both main effects and interactions. Note that there is a freely available software that provides convenient implementations: the SAFE toolbox for MATLAB [31] and SALib for Python [32].

Figure 3.20 presents the ranking of total-effect Sobol' sensitivity indices for striped bass and Atlantic menhaden. It is natural to compare these results with the PRCC-based rankings in Fig. 3.17. The total indices S_{T_i} shown in Fig. 3.20 were computed via (3.25); by construction they are nonnegative and quantify each parameter's overall contribution (including interactions) to output variance. Note that S_{T_i} indicate importance but not direction—i.e., they do not reveal whether increasing p_i tends to raise or lower the QoI.

Taken together, Figs. 3.17 and 3.20 are in qualitative agreement: parameters r and b consistently exert the greatest influence on the predator (striped bass), whereas e and d most strongly affect the prey (Atlantic menhaden).

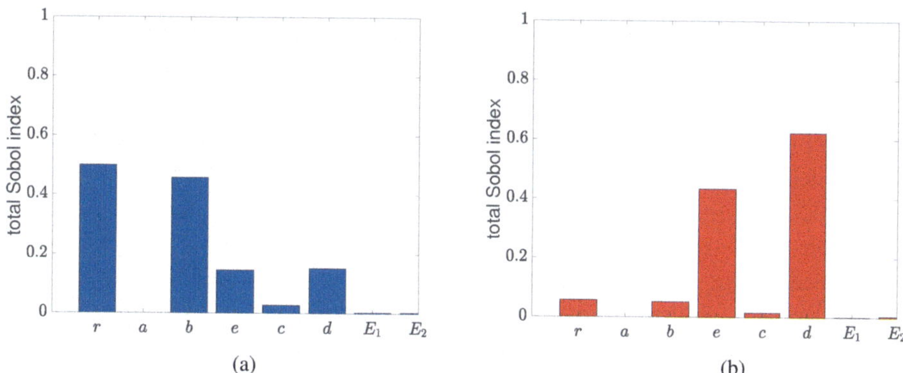

Fig. 3.20 Sobol' sensitivity ranking for striped bass the Atlantic menhaden ($t_f = 50$ years). (**a**) Striped bass. (**b**) Atlantic menhaden

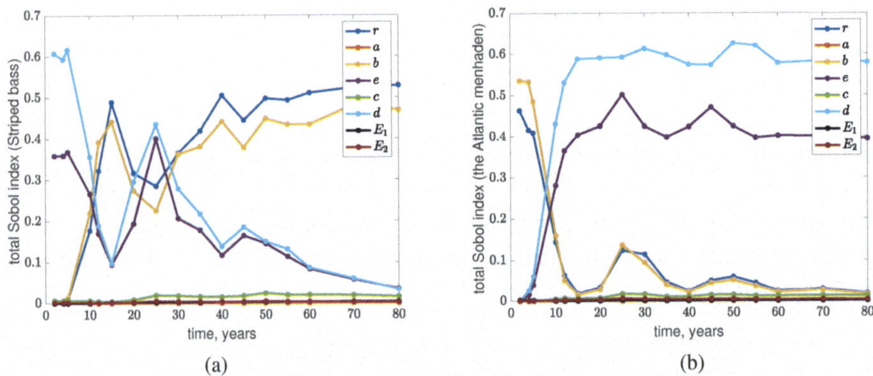

Fig. 3.21 Sobol' total sensitivity indices changing in time. (**a**) The QoI is the value of y (Striped bass). (**b**) The QoI is the value of x (Atlantic menhaden)

Some discrepancies in the subsequent rankings are apparent. We attribute these differences to limitations of PRCC under nonlinear or non-monotonic relationships between the QoI and certain parameters (see the scatter-plots in Figs. 3.12 and 3.14), whereas Sobol' indices accommodate nonlinearity and interactions.

To assess time dependence in parameter sensitivities, we computed Sobol' total-effect indices at a sequence of time points spanning calendar years 2000–2080 (equivalently, simulation times $t \in [0, t_f = 80]$). Evaluating indices at discrete times reveals how the relative importance of parameters changes during the transient toward steady state. Figure 3.21 displays the time-varying total-effect indices over the 0–80 year interval.

To further examine time-dependent sensitivities, we compare PRCC- and Sobol'-based results. For the predator (striped bass), see Fig. 3.18 (PRCC) versus Fig. 3.21a (Sobol'); for the prey (Atlantic menhaden), see Fig. 3.19 (PRCC) versus Fig. 3.21b (Sobol'). When interpreting these side by side, recall that total Sobol' indices S_{T_i} computed via (3.25) are nonnegative by definition and quantify magnitude of influence, whereas PRCC values can be positive or negative and thus convey direction as well as strength.

Both Figs. 3.18 and 3.21a show that striped bass abundance is highly sensitive to the intrinsic growth rate of Atlantic menhaden r and to the predator-on-prey effect b (impact of striped bass on menhaden). Likewise, comparing Fig. 3.19 with Fig. 3.21b highlights the roles of the striped bass natural mortality e and the prey-on-predator effect d (impact of menhaden on striped bass) in determining menhaden abundance. These results agree with the local sensitivity analysis of Panayotova et al. [21].

Regarding the next tier of influences, parameters e and d for striped bass, and r and b for menhaden, merit consideration. However, the time-resolved PRCC and Sobol' total-effect indices indicate that d and e for striped bass lose importance over time, while for menhaden the parameters r and b are primarily influential during the early stages of the process.

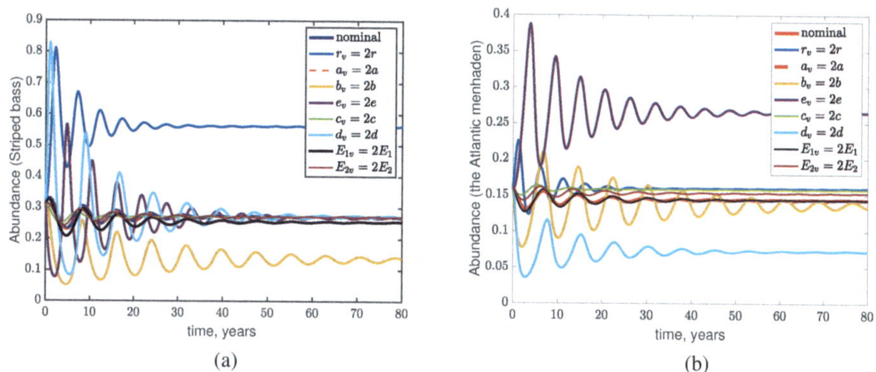

Fig. 3.22 Comparison between the solutions using the nominal and varied parameter values ($t_f =$ 80 years). (**a**) Striped bass (all parameters). (**b**) The Atlantic menhaden (all parameters)

We conclude with Fig. 3.22, which contrasts solutions obtained using nominal and varied parameter values for both striped bass and Atlantic menhaden over a time period spanning from 2000 to 2080 ($0 \leq t \leq 80$ years). These results align with the findings of the discussed sensitivity analysis methods and in reasonable agreement with the graphs by Panayotova et al. [21], depicting the rate of change of each species' population over time with respect to the system parameters. Notably, harvesting the prey exerts a stronger negative impact on the predator, whereas harvesting the predator has only a minor effect on its own steady state but yields a modest increase in the prey's steady state. This finding is consistent with the results reported by Panayotova et al. [21] and supports our previous observation that the abundance of the predator is primarily determined by the availability of prey. In addition, the prey abundance is significantly influenced by the predator's consumption effect and by reductions in predator numbers through mortality or harvest.

Important: Sobol' Method in Conclusion

The Sobol' method is among the most widely used approaches for parameter sensitivity analysis and, along with PRCC, has gained broad acceptance. A principal advantage is that Sobol' indices do not assume a monotonic relationship between the quantity of interest (QoI) and the parameters. However, convergence often requires larger sample sizes than its alternatives. Moreover, standard Sobol' estimates presume input independence; when parameters are correlated or interdependent, the indices can misattribute influence to individual parameters rather than capturing their combined effect, hindering clear discrimination. Exploratory analyses should therefore assess inter-dependencies to identify limitations and guide interpretation; techniques such as principal component analysis (PCA) can help mitigate the impact of correlated inputs [33].

(continued)

What students should retain:

- Sobol' analysis is **variance-based**: it partitions Var(z) into *main effects* (S_i) and *interaction/total effects* (S_{T_i}). Does *not* provide a sign; quantifies influence *magnitude*.
- Use Sobol' when relationships are **nonlinear and/or non-monotonic** and when **interactions** among parameters may be important—this complements rank-based PRCC, which is strong for monotonic effects but ignores interactions.
- Interpretation: S_i large $\Rightarrow$ parameter p_i explains much of the variance alone; $S_{T_i} - S_i$ measures the strength of interactions involving p_i; small S_{T_i} suggests p_i little overall influence and can often be fixed at nominal value.

Quick recipe (practical workflow):

1. Specify parameter ranges/distributions and assume inputs are **independent** (required for standard Sobol'); document the choice.
2. Choose a sampling design (Monte Carlo, LHS, or quasi–Monte Carlo). Start with $n = O(10^3)$–$O(10^4)$ model runs for moderate m and increase until indices and confidence intervals stabilize.
3. Evaluate the model, compute S_i and S_{T_i}, and report **uncertainty** (e.g. confidence intervals). Check that $\sum_i S_i \leq 1$; a large gap $1 - \sum_i S_i$ signals strong interactions.
4. Cross-check with PRCC to recover **directionality** (sign) and with scatter-plots to diagnose nonlinearity/outliers.

Software pointers: SAFE (MATLAB) and SALib (Python) provide ready-to-use implementations; use of seeds and confidence intervals is encouraged for reproducibility.

3.5 Three Species Model in the Chesapeake Bay Fishes

In this section, we discuss a three-species mathematical model that simulates the biological interactions among three important fish species in the Chesapeake Bay: the prey Atlantic Menhaden and its two competing predators: the striped bass and the blue catfish. We also considers the following ecological issues related to these three species: the overfishing of Atlantic menhaden, the invasiveness of the non-native blue catfish, and the harvesting of blue catfish as a method to control the population. This model builds naturally upon the classical Lotka-Volterra predator-prey models and allows students to apply their theoretical knowledge from previously taken classes in calculus, linear algebra, and differential equations for investigating real-world ecological problems.

3.5.1 Background

Atlantic menhaden (*Brevoortia tyrannus*) are vital to the Chesapeake Bay for both ecological and commercial reasons. Often referred to as "the most important fish in the sea," they are filter feeders that help maintain water quality by consuming phytoplankton and suspended particles. Ecologically, they serve as a critical link in the Chesapeake Bay food web, transferring energy from lower to higher trophic levels by serving as prey for a variety of predators, including striped bass, bluefish, and blue catfish. Despite their importance, the menhaden's population in the Bay experienced a sharp decline in the early 1990s and still remains low [34].

Another native species that has experienced a significant decline is the striped bass (*Morone saxatilis*)—an iconic predator in the Chesapeake Bay ecosystem and a highly valued commercial and recreational fish. Striped bass rely heavily on Atlantic menhaden as a primary food source. The decline in menhaden abundance has led to nutritional stress in some striped bass populations, contributing to reduced growth rates and increased vulnerability to disease [2]. In response to the critical decline in striped bass numbers, the Atlantic States Marine Fisheries Commission announced in 2020 new regulations aimed at reversing the downward trend in the striped bass population [35].

A growing ecological concern in the Chesapeake Bay is the impact of invasive species, particularly the blue catfish (*Ictalurus furcatus*). Introduced into Virginia tidal rivers in the early 1970s to support recreational fisheries, invasive blue catfish have expanded throughout tributaries of the Chesapeake Bay, where they now constitute up to 75% of the total fish biomass in some rivers, dramatically altering native food webs [36,37]. As novel apex predators, these invasive catfish prey on a wide range of native and anadromous species, including those of commercial and ecological importance. Their expanding populations pose a significant threat to the balance of the Bay's food web. In response, management agencies are encouraging public participation in harvesting efforts to help curb the spread of blue catfish and reduce associated ecological and economic impacts. However, the effectiveness of such a strategy in controlling the growing catfish population remains uncertain [38].

While fisheries in the Bay are traditionally managed as single-species systems, the complex, multi-species dynamics of the ecosystem must be considered when developing effective management strategies and making informed policy decisions. Mathematical modeling could be utilized to understand functional relationships among species, assess the impact of ecological interventions, and predict long-term population dynamics in the Chesapeake Bay.

Next, we develop a three-species mathematical model to investigate the relationship between the three fish species: Atlantic menhaden, striped bass, and blue catfish as illustrated in Fig. 3.23. We ignore interactions with other species, that is, we assume that there are no other predators, no other prey, and no other competitors. We also do not consider environmental or societal factors such as temperature, seasons, illegal fishing, etc. The model then serves as a tool to investigate the impact of various ecological policies on

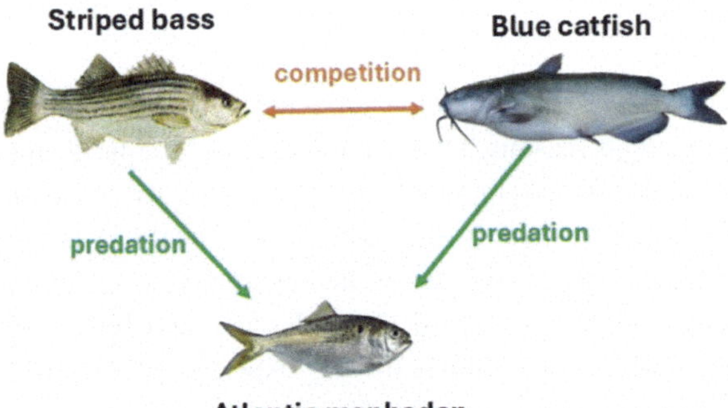

Fig. 3.23 The food web diagram between Atlantic menhaden (prey), striped bass (predator) and blue catfish (predator). The one-sided green arrows represent predation towards prey, the orange double-arrow represents competition between the two predators

long-term population dynamics of these fish and to inform strategies for their sustainable resource management.

Note that an extended version of this model using a series of modeling scenarios is published by the same authors in [39].

3.5.2 Deriving the Three-Species Model

Let $x = x(t)$ be the population of the Atlantic menhaden, $y = y(t)$ be the population of striped bass, and $z = z(t)$ be the population of blue catfish. We make the following assumptions:

1. Linear growth rate for the Atlantic menhaden (x).
2. The two predators, striped bass (y) and catfish (z) are in competition, both preying on the Atlantic menhaden.
3. Blue catfish (z) is an invasive species.
4. Striped bass (y) depends on the Atlantic menhaden (x) but catfish (z) does not.

The system of differential equations describing the three-species biological interactions under the above assumptions is given by:

$$\begin{cases} \dfrac{dx}{dt} = ax - bxy - cxz \\ \dfrac{dy}{dt} = -dy + bxy - \alpha yz \\ \dfrac{dz}{dt} = ez + cxz - \beta yz \end{cases} \tag{3.26}$$

Table 3.6 The functions and parameters in the model (3.26). All parameters are positive

Parameter	Description
$x(t)$	Population of prey (Atlantic menhaden)
$y(t)$	Population of predator (striped bass)
$z(t)$	Population of invasive predator (blue catfish)
a	Growth rate parameter of x
b	Preying effect of y on x
c	Preying effect of z on x
d	Death rate parameter of the predator y in the absence of prey x
e	Growth rate parameter of z
h_1	Harvesting effect on x
h_2	Harvesting effect on z
α	Competition effect of z on y
β	Competition effect of y on z

Note that there is unequal competition between the predators, i.e. $\alpha \neq \beta$. When the catfish is more aggressive than the striped bass then $\alpha > \beta > 0$ and when the striped bass is more aggressive than the catfish then $0 < \alpha < \beta$. The case when $\alpha = \beta$ is left as a practice Problem 3.15 for the students. Table 3.6 summarizes functions and parameters in the model (3.26).

3.5.3 Equilibria and Their Biological Relevance

Setting the derivatives equal to zero yields four boundary equilibrium points and one interior equilibrium point for (3.26):

$$Q_0 = (0, 0, 0),\ Q_1 = \left(0, \frac{e}{\beta}, \frac{-d}{\alpha}\right),\ Q_2 = \left(\frac{-e}{c}, 0, \frac{a}{c}\right),\ Q_3 = \left(\frac{d}{b}, \frac{a}{b}, 0\right), \tag{3.27}$$

and

$$Q_4 = \left(\frac{(cd + \alpha a)\beta - \alpha be}{(\alpha + \beta)bc}, \frac{cd + be + \alpha a}{(\alpha + \beta)b}, \frac{a\beta - (cd + be)}{(\alpha + \beta)c}\right). \tag{3.28}$$

Note that Q_1 and Q_2 have negative components and are not biologically possible. For the equilibrium point Q_4 to be biologically relevant, all of its components must be positive. Observe that its y-component is always positive while its z-component is positive provided

$$a\beta > cd + be \text{ or } \beta > \frac{cd + be}{a}.$$

Let us denote $\beta_{crit} = \dfrac{cd+be}{a}$. If $\beta > \beta_{crit}$ the z-component of Q_4 will be positive. Let us prove that the condition $\beta > \beta_{crit}$ is sufficient for the x-component of Q_4 to be also positive. Using this condition, substitute β in the numerator of the x-component as follows

$$(cd + a\alpha)\beta > \frac{(cd + a\alpha)(cd + be)}{a}.$$

The right-hand side of this inequality can be simplified to

$$\frac{(cd)^2 + cdbe + a\alpha cd + a\alpha be}{a} = \frac{(cd)^2 + cdbe + a\alpha cd}{a} + \alpha be > \alpha be.$$

Thus, $(cd + a\alpha)\beta > \alpha be$ so the first component of Q_4 is positive. In conclusion, if β is larger than some critical value β_{crit}, defined by the system's parameters, then there exists an interior equilibrium point. Biologically, this means that co-existence of the three species is possible for some period of time. Also observe that when $\alpha = \beta = \beta_{crit}$, the equilibrium Q_4 reduces to the equilibrium Q_3. In this case, we say that β_{crit} is a bifurcation value. Let us summarize our results into a theorem:

Theorem 3.1 *Suppose* $\beta > \beta_{crit} = \dfrac{cd+be}{a}$. *Then the system (3.26) has three biologically relevant equilibria, namely,* $Q_0 = (0, 0, 0)$, $Q_3 = (d/b, a/b, 0)$, *and*

$$Q_4 = \left(\frac{(cd + \alpha a)\beta - \alpha be}{(\alpha + \beta)bc}, \frac{cd + be + \alpha a}{(\alpha + \beta)b}, \frac{a\beta - (cd + be)}{(\alpha + \beta)c} \right).$$

In the next section, we will analyze the stability of these three equilibria using linear stability analysis.

3.5.4 Elementary Stability Theory of Dynamical Systems

We will first start by reminding some basic definitions and results from dynamical systems which will be then used to analyze the stability of the three equilibria Q_0, Q_3, and Q_4.

Definition 3.11 An equilibrium point $\bar{w} = (\bar{x}, \bar{y}, \bar{z})$ is called **stable** if a solution $w(t) = (x(t), y(t), z(t))$ based nearby remains close to $\bar{w}$ for all time. Otherwise, the point $\bar{w}$ is called an unstable equilibrium point. An equilibrium point is (locally) **asymptotically stable** if it is stable and the state of the system converges to the equilibrium point as time increases.

In order to classify the stability of our equilibria, the definition requires that we characterize the behavior of solutions near our equilibria. This is accomplished by linearizing our system (3.26). Students have been introduced to the concept of linearization from their Differential Calculus courses; they know that the tangent line provides to a nonlinear function at a point in its domain a good approximation of the graph of the function. Analogously, the behavior of nonlinear systems such as (3.26) near the equilibrium point is similar to the behavior of linear systems near that equilibrium point. This is the so-called Hartman-Grobman Theorem and we refer to [40–42] for proof and the precise statement.

Towards this end, we consider the linearized system

$$\dot{w}(t) = J(\bar{w})w, \tag{3.29}$$

where $w = (x(t), y(t), z(t))$ and $J(w)$ is the 3-by-3 Jacobian matrix given by

$$J(x, y, z) = \begin{pmatrix} a - by - cz & -bx & -cx \\ by & bx - d - \alpha z & -\alpha y \\ cz & -\beta z & e + cx - \beta y \end{pmatrix}. \tag{3.30}$$

The entries of the Jacobian matrix $J = (a_{ij})$ are obtained by taking the partial derivatives,

$$a_{ij} = \frac{\partial f_i}{\partial x_j}, \text{ where } (x_1, x_2, x_3) = (x, y, z).$$

For example,

$$a_{11} = \frac{\partial f_1}{\partial x} = a - by - cz, \quad a_{12} = \frac{\partial f_1}{\partial y} = -bx, \quad a_{13} = \frac{\partial f_1}{\partial z} = -cx.$$

When $J(\bar{w})$ has no eigenvalues with zero real part, then $\bar{w}$ is called a **hyperbolic equilibrium point** and the asymptotic behavior of solutions near $\bar{w}$ is determined by the linearization (3.29). If any one of the eigenvalues of $J(\bar{w})$ has a zero real part, then interesting behaviors may occur. Linear stability analysis is accomplished via a fundamental theorem [40–42]:

Theorem 3.2 *An equilibrium point $\bar{w}$ is asymptotically stable if all the eigenvalues of $J(\bar{w})$ have negative real parts. If at least one of the eigenvalues of $J(\bar{w})$ has a positive real part, then the equilibrium point is unstable.*

Suppose that the three eigenvalues of $J(\bar{w})$ are $\lambda_1 < 0, \lambda_2 = 0$, and $\lambda_3 > 0$ with corresponding eigenvector v_1, v_2, v_3. Solutions which start close to the equilibrium point $\bar{w}$ will decay in the direction of v_1 and will grow in the direction of v_3. Since $\lambda_2 = 0$, we cannot conclude from linearization whether the solutions decay or grow in the direction of v_2. The space spanned by the eigenvector v_1, v_3 is called the local stable and local unstable manifolds of $\bar{w}$, respectively; the space spanned by the eigenvector v_2 is called the local center manifold of $\bar{w}$. At this point, a manifold may be briefly described as a generalization of a linear subspace of $\mathbb{R}^n$. Points, lines, planes, arcs, and spheres are examples of manifolds. Students who become interested in this topic may want to look into the area of mathematics called Differential Geometry.

To make a direct conclusion about the stability of an equilibrium, one has to analyze the nature of the eigenvalues, that is, the roots of the characteristic equation

$$det(J(\bar{w}) - \lambda I) = 0.$$

In some cases, finding the roots of the characteristic equation may be obtained by hand. Indeed, for the system that we are considering in this project, the characteristic equation is a third-degree polynomial with real coefficients. In case solving the roots is not possible, then there are some analytic, algebraic, and computational options. In the next section, we will demonstrate one of these options and the other two are left as practice problems.

The following theorem summarizes the existence of **stable**, **center**, and **unstable** manifolds for a three-dimensional system and is oftentimes referred to as the Center Manifold Theorem. For the precise statement and proof, we refer to [40–42]:

Theorem 3.3 *Suppose f is a smooth vector field in $\mathbb{R}^3$. Assume that the system $\dot{w} = f(w)$ has an equilibrium point $\bar{w}$ and its linearization at $\bar{w}$ is given by $\dot{w} = J(\bar{w})w$. Suppose σ_s, σ_c, σ_u, contain the negative, zero, and positive eigenvalues of $J(\bar{w})$, respectively. Let E^s, E^c, E^u be the eigenspaces associated with σ_s, σ_c, σ_u, respectively. Then there exist smooth stable W^s, center W^c, and unstable W^u manifolds tangent to E^s, E^c, and E^u, respectively at $\bar{w}$. The stable and unstable manifolds are unique, but W^c need not be.*

3.5.5 Stability of Biologically Relevant Equilibria

We have seen that there are three biologically relevant equilibrium points. Let us call Q_0 the trivial equilibrium point, Q_3 the catfish-free boundary equilibrium point, and Q_4 the interior co-existence equilibrium point. In this section, we will show that they are all unstable.

Stability Analysis for $Q_0 = (0, 0, 0)$

The Jacobian matrix $J(Q_0) = J(0, 0, 0)$ is given by

$$J(0,0,0) = \begin{pmatrix} a & 0 & 0 \\ 0 & -d & 0 \\ 0 & 0 & e \end{pmatrix}$$

with eigenvalues $a > 0, -d < 0, e > 0$ and with corresponding eigenvectors $(1, 0, 0)$, $(0, 1, 0)$, and $(0, 0, 1)$. By Theorem 3.3, we know that the negative eigenvalue has a corresponding one-dimensional stable manifold, or curve, tangent to the eigenvector $(0, 1, 0)$ at the equilibrium $(0, 0, 0)$. This stable curve is actually the y-axis which means that $y(t) \to 0$ as $t \to \infty$; this is biologically relevant as in the absence of x , the prey, the predator y will become extinct. Corresponding to the two positive eigenvalues, there is a two-dimensional unstable manifold or a plane, tangent at the equilibrium $(0, 0, 0)$ and spanned by the vectors $(1, 0, 0)$ and $(0, 0, 1)$. This two-dimensional unstable manifold is the xz-plane which from biological perspective means that both the prey and the invader will grow without a boundary in the absence of y, i.e. $x(t) \to \infty$ and $z(t) \to \infty$.

Since $J(Q_0)$ has two positive eigenvalues and one negative eigenvalue, Q_0 is unstable and is called a hyperbolic equilibrium point. Moreover, in this case, we say that Q_0 is a hyperbolic saddle point.

Stability Analysis for $Q_3 = \left(\frac{d}{b}, \frac{a}{b}, 0\right)$

The equilibrium $Q_3 = \left(\dfrac{d}{b}, \dfrac{a}{b}, 0\right)$ is an unstable non-hyperbolic equilibrium point. To see this, the Jacobian at E_3 is given by the following matrix

$$J\left(\frac{d}{b}, \frac{a}{b}, 0\right) = \begin{pmatrix} 0 & -d & \dfrac{-cd}{b} \\ a & 0 & \dfrac{-\alpha a}{b} \\ 0 & 0 & \dfrac{cd + be - \beta a}{b} \end{pmatrix},$$

with characteristic equation $g(x) = (x - \lambda_s)(x^2 + ad)$ where $\lambda_s = \dfrac{cd + be - \beta a}{b}$. The three eigenvalues are λ_s and the two purely imaginary eigenvalues $\pm i\sqrt{ad}$.

If $\beta > \beta_{crit}$, observe that λ_s is negative and hence by Theorem 3.3, there is a one-dimensional stable manifold at Q_3 tangent to the eigenvector spanned by

$$(\beta_1\beta_3 - \beta_2 d, a\beta_1 + \beta_2\beta_3, \beta_3^2 + ad),$$

where $\beta_1 = \frac{-cd}{b}$, $\beta_2 = \frac{-\alpha a}{b}$, and $\beta_3 = \lambda_s$.

Corresponding to the purely imaginary eigenvalues $\pm i\sqrt{ad}$, there exists a two-dimensional center manifold at Q_3 tangent to the space spanned by the eigenvectors

$$\begin{pmatrix} 0 \\ 1 \\ 0 \end{pmatrix} \pm \begin{pmatrix} 1 \\ 0 \\ 0 \end{pmatrix} \sqrt{\frac{d}{a}}\, i.$$

The eigenspace spanned by these vectors is the xy-plane. As not all eigenvalues have negative real parts, this equilibrium is unstable.

Stability Analysis for Q_4

Next, we will examine the stability of the equilibrium $Q_4 = (x^*, y^*, z^*)$, where

$$x^* = \frac{(cd + \alpha a)\beta - \alpha be}{(\alpha + \beta)bc}, \; y^* = \frac{cd + be + \alpha a}{(\alpha + \beta)b}, \; z^* = \frac{\beta a - (cd + be)}{(\alpha + \beta)c}.$$

Recall that we assume $\beta > \beta_{crit}$ so that x^*, y^*, z^* are always positive. The Jacobian matrix (3.30) at this point is given by

$$J(x^*, y^*, z^*) = \begin{pmatrix} 0 & -bx^* & -cx^* \\ by^* & 0 & -\alpha y^* \\ cz^* & -\beta z^* & 0 \end{pmatrix}.$$

where (x^*, y^*, z^*) is the solution to the system

$$\begin{cases} a - by - cz = 0 \\ bx - d - \alpha z = 0 \\ cx + e - \beta y = 0. \end{cases}$$

To find the eigenvalues λ, we need to analyze the roots of the characteristic equation

$$\lambda^3 + (b^2x^*y^* + c^2x^*z^* - \alpha\beta y^*z^*)\lambda - bc(\alpha + \beta)x^*y^*z^* = 0,$$

that is,

$$\lambda^3 + B\lambda + C = 0, \tag{3.31}$$

where $B = b^2x^*y^* + c^2x^*z^* - \alpha\beta y^*z^*$ and $C = -bc(\alpha + \beta)x^*y^*z^*$.

Solving for the closed-form of the roots of (3.31) is not a straightforward computation. Here, to analyze the roots of this polynomial we use the Routh-Hurwitz Criteria. This theorem gives a necessary and sufficient condition for a polynomial to have all zeros with negative real parts [43]. Here, we include this theorem's version for the particular case of a third-degree polynomial:

Theorem 3.4 (Routh-Hurwitz Criteria) *For a third-degree polynomial of the form $\lambda^3 + A\lambda^2 + B\lambda + C = 0$, the three roots will have negative real parts if and only if the following three conditions are satisfied*

$$A, C > 0 \text{ and } AB - C > 0.$$

From this theorem follows that if at least one of the coefficients A or C is not strictly positive, which is our case, then it is not possible for all three roots to have negative real parts and hence the equilibrium point Q_4 is unstable.

Summarizing our results on the behavior of the equilibria of the system (3.26), we have

Theorem 3.5 *Suppose* $\beta > \beta_{crit} = \dfrac{cd + be}{a}$. *Then the system (3.26) has the following biologically relevant equilibria:*

$$Q_0 = (0, 0, 0), \; Q_3 = \left(\frac{d}{b}, \frac{a}{b}, 0\right), \text{ and}$$

$$Q_4 = \left(\frac{(cd + \alpha a)\beta - \alpha be}{(\alpha + \beta)bc}, \frac{cd + be + \alpha a}{(\alpha + \beta)b}, \frac{a\beta - (cd + be)}{(\alpha + \beta)c}\right),$$

which are all unstable.

3.6 Three-Species Model with Harvesting

In this section we use the three-species model (3.26) describing interactions between the Atlantic menhaden, striped bass, and blue catfish to investigate the impact of harvesting on long-term population dynamics of these fish. We study the overfishing of Atlantic menhaden and the harvesting of blue catfish as a method to control the population.

3.6.1 Overfishing of Menhaden

Any fish management policy must consider the interactions between species when deciding on harvesting quotes, as the population dynamics between species is inevitably linked. Only recently, the Atlantic States Marine Fisheries Commission (ASMFC) has taken the first step to formally consider the importance of one species to others, particularly the importance of Atlantic menhaden to other predators, including the striped bass. In their new management policy, they committed to including ecological reference points on menhaden's population and reduced its harvesting quota by 10% for 2021 and 2022 fishing seasons [44].

We start our analysis by modifying the general model (3.26) to include harvesting of the prey (Atlantic menhaden) $x(t)$, assuming that it occurs at a linear rate, $h_1 x$, for some positive constant h_1. Biologically this means that the harvesting of the prey happens in direct proportion with its growth. Additionally, we also assume that the predators striped bass and blue catfish are equally competitive, i.e. $\alpha = \beta$. The new model is given as follows:

$$\begin{cases} \dfrac{dx}{dt} = ax - bxy - cxz - h_1 x = Ax - bxy - cxz \\ \dfrac{dy}{dt} = bxy - dy - \alpha yz \\ \dfrac{dz}{dt} = ez + cxz - \alpha yz. \end{cases} \tag{3.32}$$

When harvesting of the prey is accomplished at a higher rate than the prey's growth, we say that overfishing occurs. Defining $A = a - h_1$, we see that overfishing happens when $A < 0$.

Equilibria

As before, setting the derivatives equal to zero in (3.32) solves the equilibrium points. We obtain

$$Q_0 = (0, 0, 0),\ Q_1 = \left(0, \frac{e}{\alpha}, \frac{-d}{\alpha}\right),\ Q_2 = \left(\frac{-e}{c}, 0, \frac{a - h_1}{c}\right),\ Q_3 = \left(\frac{d}{b}, \frac{a - h_1}{b}, 0\right),$$

and the co-existence equilibrium

$$Q_4 = \left(\frac{cd - be + \alpha(a - h_1)}{2bc}, \frac{cd + be + \alpha(a - h_1)}{2\alpha b}, \frac{\alpha(a - h_1) - (cd + be)}{2\alpha c}\right).$$

Let us observe that Q_1, Q_2, and Q_3 and Q_4 are not biologically relevant as some of their components are negative due to the assumption $A = a - h_1 < 0$. In particular, when overfishing of Atlantic menhaden happens, the interior co-existence equilibrium point is not biologically possible.

Theorem 3.6 *Suppose there is overfishing of the prey $h_1 > a$. Then the system (3.32) has only one biologically relevant equilibrium point, namely the trivial equilibrium $Q_0 = (0, 0, 0)$.*

Linearization and Stability Analysis

The linearization of (3.32) is represented by the matrix

$$J(x, y, z) = \begin{pmatrix} a - by - cz - h_1 & -bx & -cx \\ by & bx - d - \alpha z & -\alpha y \\ cz & -\alpha z & cx + e - \alpha y \end{pmatrix}.$$

Let us now analyze the behavior of the system at the only biologically relevant equilibrium point, Q_0. Observe that

$$J(0, 0, 0) = \begin{pmatrix} a - h_1 & 0 & 0 \\ 0 & -d & 0 \\ 0 & 0 & e \end{pmatrix}$$

with eigenvalues $a - h_1 < 0, -d < 0, e > 0$ and the corresponding eigenvectors are $(1, 0, 0)$, $(0, 1, 0)$, and $(0, 0, 1)$. Since we assume that $h_1 > a$, we have two negative eigenvalues and one positive eigenvalue. Thus, Q_0 is unstable. Moreover, using

Theorem 3.3, there is an invariant two-dimensional stable manifold that is tangent to the space spanned by eigenvectors $(1, 0, 0)$ and $(0, 1, 0)$, which is the xy-plane. On the xy-plane, solutions approach the equilibrium $Q_0 = (0, 0, 0)$, meaning that $x(t) \to 0$ as $t \to \infty$ and $y(t) \to 0$ as $t \to \infty$. This is biologically relevant as if the prey is overfished, then in time both the prey and the predator will become extinct. In addition, there is also an one-dimensional unstable manifold spanned by the eigenvector $(0, 0, 1)$, or the z-axis, on which $z(t) \to \infty$ or the catfish will grow without limit.

3.6.2 Harvesting of Blue Catfish

Recently, fisheries management strategies have been developed by NOAA and its local partner agencies to examine the potential impacts of commercial and recreational harvesting of the blue catfish in the Chesapeake Bay, as "a means to reduce their abundance and mitigate their range expansion and ecological impacts" [45]. To curb their spread, NOAA has recommended several control measures, including consumer-oriented initiatives that promote the harvesting and consumption of this invasive species [46].

In this section, we would like to use our mathematical model to explore the inquiry: will harvesting of the catfish solve its invasive effect on the Chesapeake Bay? In other words, does it help to *eat the invaders* [46]?

Let us modify the model (3.26) to include harvesting of the catfish. Assume that harvesting of the catfish occurs at a linear rate $h_2 z$, for some positive constant h_2, that is, the harvesting of z happens in direct proportion to its population. Additionally, as before, we also assume that the two predators are equally competitive by setting $\alpha = \beta$. We get the system

$$\begin{cases} \dfrac{dx}{dt} = ax - bxy - cxz \\ \dfrac{dy}{dt} = -dy + bxy - \alpha yz \\ \dfrac{dz}{dt} = ez + cxz - \alpha yz - h_2 z = Ez + cxz - \alpha yz, \end{cases} \tag{3.33}$$

where $E = e - h_2$. Note that unlike the other parameters, E may be positive, negative, or zero.

Equilibrium Points

Setting the derivatives in (3.33) equal to zero, we find that there are four boundary equilibrium points:

$$Q_0 = (0, 0, 0),\ Q_1 = \left(0, \frac{E}{\alpha}, \frac{-d}{\alpha}\right),\ Q_2 = \left(\frac{-E}{c}, 0, \frac{a}{c}\right),\ Q_3 = \left(\frac{d}{b}, \frac{a}{b}, 0\right),$$

and there is one interior co-existence equilibrium point:

$$Q_4 = \left(\frac{cd - bE + \alpha a}{2bc}, \frac{cd + bE + \alpha a}{2\alpha b}, \frac{\alpha a - (cd + bE)}{2\alpha c} \right).$$

For the equilibrium Q_4 to exist, we need to require that its components are positive:

$$\begin{cases} cd - bE + \alpha a & > 0 \\ cd + bE + \alpha a & > 0 \\ \alpha a - (cd + bE) > 0. \end{cases}$$

Now, let us study parameter intervals using these inequalities. Manipulation of the inequalities depends on the parameter that one wishes to investigate. In this case we want to find a bound for the newly defined parameter E. Toward this end, let us analyze biological relevance of the found equilibria with respect to the sign of $E = e - h_2$, where e is the growth rate and h_2 is the harvesting parameter of the catfish.

Case 1 $E > 0$. This is the case when $h_2 < e$ or when the rate of harvesting does not exceed the growth rate of the species z. We shall call this case (a plain) harvesting. Observe that equilibria Q_1 and Q_2 are not biologically relevant as they have a negative component. The equilibrium Q_4 would exist if its all components are positive. Note that its y-component is always positive so we only need to require

$$\begin{cases} \alpha a + cd > bE \\ \alpha a - cd > bE. \end{cases} \tag{3.34}$$

Expressing E from here into one inequality gives

$$0 < E < \frac{\alpha a - cd}{b} < \frac{\alpha a + cd}{b},$$

if $\alpha > \alpha_{crit} = \dfrac{cd}{a}$. Equivalently, it is sufficient to require that

$$0 < E < \frac{\alpha a - cd}{b} \quad \text{and} \quad \alpha > \alpha_{crit} = \frac{cd}{a}. \tag{3.35}$$

Hence, the equilibrium point E_4 exists when the competition is sufficiently high.

Case 2 $E < 0$. This is the case when $h_2 > e$ or when the rate of harvesting exceeds the growth rate of the species z. We shall call this the **overfishing** of the catfish population. In this case Q_2 is a biologically relevant equilibrium as well as Q_0 and Q_3. For the Q_4 we see that its x-component is always positive, so we need to require only that

$$\begin{cases} \alpha a + cd > -bE \\ \alpha a - cd > -bE. \end{cases} \tag{3.36}$$

Keeping in mind that $-E > 0$, these reduces to the same inequalities (3.35), i.e

$$0 < |E| < \frac{\alpha a - cd}{b} \quad \text{and} \quad \alpha > \alpha_{crit} = \frac{cd}{a}. \tag{3.37}$$

Hence, as before, the equilibrium Q_4 is biologically relevant with overfishing of the catfish as long as the competition parameter is sufficiently large value. If the parameter α is not big enough, then the equilibrium Q_4 will not have positive components and hence Q_4 will not be biologically relevant. This means that even when there is overfishing of the catfish, all three species may not co-exist. If the competition strength between the striped bass and the catfish is below the critical value $\alpha_{crit} = \frac{cd}{a}$, then the three species will **not co-exist**. Hence, when $E < 0$, there are three boundary equilibrium points, Q_0, Q_2, Q_3 and one interior co-existence equilibrium point Q_4.

Case 3 $E = 0$. This is the case when the harvesting rate is exactly equal to the growth rate. Then the equilibrium Q_4 exists, that is, it will have positive components provided $-cd + \alpha a > 0$ or $\alpha > \alpha_{crit} = \dfrac{cd}{a}$. Note that this is a bifurcation value for α. When $E = 0$, the equilibrium points are Q_0, Q_2, Q_3, and Q_4.

These results on the equilibria are summarized in the following theorem:

Theorem 3.7 *Suppose* $|E| < E_{crit} = \dfrac{\alpha a - cd}{b}$ *and* $\alpha > \alpha_{crit} = \dfrac{cd}{a}$, *where* $E = e - h_2$. *The system (3.33) has five equilibrium points:*

$$Q_0 = (0, 0, 0),\ Q_1 = \left(0, \frac{E}{\alpha}, \frac{-d}{\alpha}\right),\ Q_2 = \left(\frac{-E}{c}, 0, \frac{a}{c}\right),\ Q_3 = \left(\frac{d}{b}, \frac{a}{b}, 0\right),$$

and

$$Q_4 = \left(\frac{cd - bE + \alpha a}{2bc}, \frac{cd + bE + \alpha a}{2\alpha b}, \frac{\alpha a - (cd + bE)}{2\alpha c}\right).$$

1. *If* $E \leq 0$ *then* Q_0, Q_2, Q_3, Q_4 *are biologically relevant.*
2. *If* $E > 0$ *then* Q_0, Q_3, Q_4 *are biologically relevant.*

In particular, the interior co-existence equilibrium point Q_4 *exists for both positive and negative* E, *as long as its absolute value is less than a critical value* E_{crit}.

This theorem highlights an interesting result that when harvesting on the invasive species happens, competition between the two predators must be high enough in order for the three species to co-exist. Competition, in this case, is essential and beneficial to the ecological dynamics of this system.

Linearization and Stability Analysis for (3.33)

The linearization is given by the 3-by-3 matrix $\dot{w}(t) = J(\bar{w})w$ where $w = (x(t), y(t), z(t))$ and $J(w)$ is the 3-by-3 Jacobian matrix given by

$$J(x, y, z) = \begin{pmatrix} a - by - cz & -bx & -cx \\ by & bx - d - \alpha z & -\alpha y \\ cz & -\alpha z & E + cx - \alpha y \end{pmatrix}.$$

As described and demonstrated in the previous section, stability analysis requires analyzing the roots of the resulting characteristic equation for $J(\bar{w})$ where $\bar{w} \in \{Q_0, Q_2, Q_3, Q_4\}$. Here, we assume that the competition parameter α and the harvesting satisfy conditions (3.37). A summary of the equilibria and their stability is presented in Table 3.7.

In conclusion, we can say that the policy that recommends eating the catfish to control its invasive growth [46] may not be able to accomplish the following: drive the catfish to extinction, eradicate competition with the striped bass, or affect the menhaden's population growth adversely; however, mathematical analysis shows that overfishing the catfish may allow all three species to coexist.

3.7 Long-Term Population Dynamics of the Three Fish Species

In the previous sections, we employed theoretical analysis to examine the local stability of the equilibrium solutions for the three-species model, thereby gaining insight into the short-term behavior of the system in the vicinity of these equilibria. In this section, we shift our focus toward the long-term population dynamics of the three fish species by conducting numerical simulations of the mathematical model (3.26). These simulations will allow us to explore scenarios beyond the analytical reach of stability analysis, capturing transient behaviors, potential oscillations, and asymptotic trends over extended time periods. All numerical computations and visualizations are carried out using MATLAB and are based on simulations performed in [39].

3.7.1 Exploration via Numerical Time-Evolutions

Figure 3.24 shows the long-term behavior of the fish populations for some particular values of the parameters a, b, c, d, e given in Table 3.8 with initial conditions (100, 10, 10). Note

Table 3.7 The table presents a summary of the equilibria of model (3.33) and their stability when harvesting of the catfish happens, where $E = e - h_2$

Equilibrium point	Characteristic equation	Eigenvalues	Stability result
$Q_0(0, 0, 0)$	$(\lambda - a)(\lambda + d)(\lambda - E) = 0$	$a, -d, E$	This equilibrium point exists and is unstable for all values of E
$Q_2\left(\frac{-E}{c}, 0, \frac{a}{c}\right)$	$(\lambda + \lambda_{t+})(\lambda^2 - aE) = 0$, $\lambda_{t+} = \frac{bE + cd + \alpha a}{c}$	$\pm\sqrt{aE}$ and $-\lambda_{t+}$	$E > 0$: **not an equilibrium** $E = 0$: **unstable** $E < 0$: **unstable**
$Q_3\left(\frac{d}{b}, \frac{a}{b}, 0\right)$	$(\lambda - \lambda_{t-})(\lambda^2 + ad) = 0$, $\lambda_{t-} = \frac{bE + cd - \alpha a}{b}$	$\pm i\sqrt{ad}$ and λ_{t-}	This equilibrium point exists and is unstable for all values of E
$Q_4(x^*, y^*, z^*)$ $x^* = \frac{cd - bE + \alpha a}{2bc}$, $y^* = \frac{cd + bE + \alpha a}{2\alpha b}$, $z^* = \frac{\alpha a - (cd + bE)}{2\alpha c}$	$\lambda^3 + B\lambda + C = 0$, $B = b^2x^*y^* + c^2x^*z^* - \alpha^2y^*z^*$, $C = -2bc\alpha x^*y^*z^*$	Computing the closed-form of roots is not straightforward	This equilibrium point exists and is unstable for all values of E

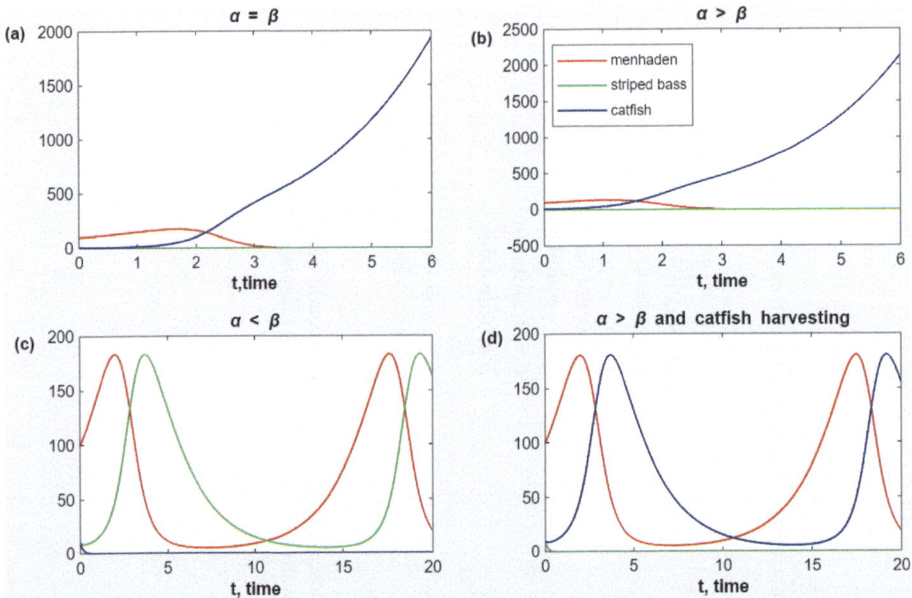

Fig. 3.24 Menhaden, striped bass, and catfish distributions over time for the following values of the parameters: $a = d = e = 0.5$, $b = c = 0.01$ with initial conditions (100, 10, 10). Panels (**a**) $\alpha = \beta = 0.5$, (**b**) $\alpha = 0.8$ and $\beta = 0.2$, (**c**) $\alpha = 0.2$ and $\beta = 0.8$, (**d**) $\alpha = 0.8$, $\beta = 0.2$ and harvesting parameter of the catfish is $h_2 = 1$

Table 3.8 The parameter values used in the simulation are summarized in this table

Parameter	Description	Value
a	Growth rate parameter of x	0.5
b	Preying effect of y on x	0.01
c	Preying effect of z on x	0.01
d	Death rate of the predator y in the absence of prey x	0.5
e	Growth rate parameter of z	0.5

that these values are chosen randomly and are not based on observational data. Panel (a) illustrates the populations when $\alpha = \beta$ and the two predators are equally competitive. The catfish (represented by the blue curve) grows without bound. The other two species die out: the striped bass (represented by the green curve) dies out earlier than the menhaden (represented by the red curve). This behavior among the three species is still reflected in the case when $\alpha > \beta$, that is, the catfish is a stronger competitor than the striped bass as shown in panel (b). If we assume that the catfish is less competitive or less aggressive than the striped bass $\alpha < \beta$ then as shown in panel (c), the catfish becomes extinct while the menhaden and the striped bass co-exist. Furthermore, in this case, we observe periodic shifted behavior between menhaden and bass that is typically found in two-species predator-prey models.

However, in reality the blue catfish is much more competitive in the following sense. According to [47], blue catfish constitute up to 75% of the total fish biomass in some tidal rivers of Virginia, particularly parts of the James and Rappahannock rivers. Moreover, blue catfish prey on a wide range of species, including juvenile striped bass, and not only on Atlantic menhaden. Hence, it is biologically more realistic to assume that $\alpha > \beta$. Our model (see Fig. 3.24b) predicts that ultimately, the blue catfish grows in an unbounded manner compared with the native species, the striped bass and the Atlantic menhaden. If no serious measures are taken to prevent the wide spread of the invasive blue catfish, the native species will be seriously affected and may even become extinct. Panel (d) in Fig. 3.24 presents a simulation of the time-evolution of the three species' populations when there is harvesting of the catfish with the more realistic case of $\alpha > \beta$. If the harvesting term is assumed to be a linear rate $h_2 z$, and the harvesting parameter h_2 is high enough, then it is possible for the catfish and the Atlantic menhaden to co-exist, and they exhibit the typical predator-prey phase-shifted periodic behavior. However, the striped bass dies out.

3.7.2 Further Numerical Investigations: Harvesting of the Blue Catfish

Continuing with the more biologically realistic assumption that $\alpha > \beta$, Fig. 3.25 shows the effect of harvesting on the catfish. Here, we would like to investigate if harvesting is able

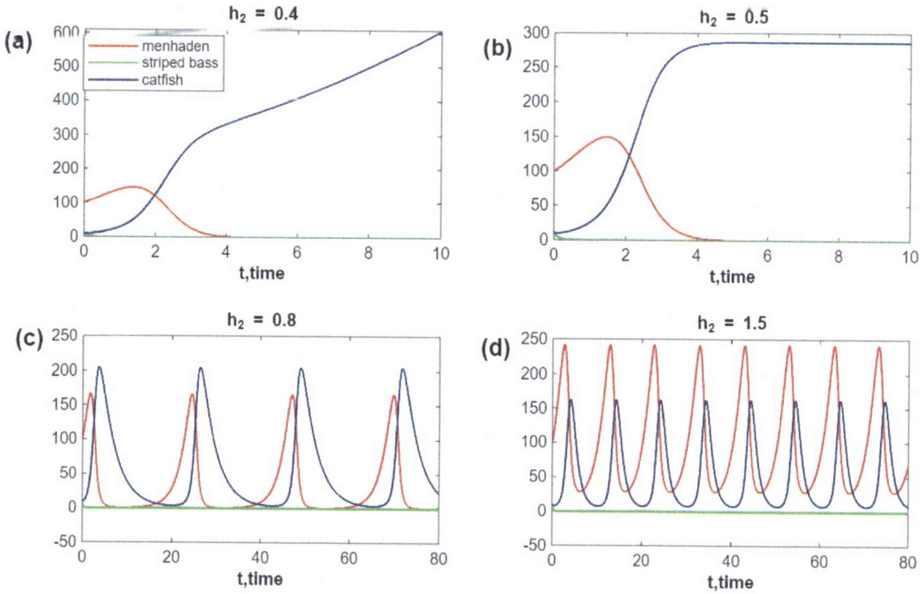

Fig. 3.25 Time-evolutions on the population of menhaden, striped bass, and catfish when there is harvesting of catfish. The following values of the parameters were used: $a = d = e = 0.5$, $b = c = 0.01$, $\alpha = 0.8$, $\beta = 0.2$ with initial conditions (100, 10, 10). Panels (**a**) $h_2 = 0.4$, (**b**) $h_2 = 0.5$, (**c**) $h_2 = 0.8$, (**d**) $h_2 = 1.5$

to control the invasive effect of the catfish on the Chesapeake Bay. In these simulations, we will keep $e = 0.5$ constant, where e is the growth parameter for the catfish.

Panel (a) shows that a harvesting rate $h_2 = 0.4$ is not sufficient to control the population growth of the catfish. Panel (b) shows that with a harvesting rate of $h_2 = 0.5 = e$, the (blue) population curve for the catfish exhibits bounded growth. Thus, we say that in this case, harvesting is able to control the growth of the catfish. Unfortunately, the menhaden eventually dies out. The second predator (the striped bass) dies even before the menhaden, due to the decreasing population of its prey and the aggressive behavior of its competitor.

Now, what happens when the harvesting rate of the catfish exceeds its growth rate, that is, when overfishing of the catfish happens? Panels (c) and (d) show two cases, with $h_2 = 0.8$ and $h_2 = 1.5$ respectively.

Panel (c) shows that with aggressive harvesting $h_2 = 0.8 > 0.5 = e$, a periodic phase-shifted behavior becomes evident for the catfish and its menhaden prey. The striped bass, which is a second predator and also a competitor to the catfish, dies out.

Panel (d) shows that with even more aggressive harvesting $h_2 = 1.5 > 0.5 = e$, a periodic phase-shifted behavior also occurs between the predator and prey. The striped bass also dies out in this case. Hence, we can conclude that it is possible to control the exponential growth of the invasive species with aggressive harvesting.

Finally, observe that between panels (c) and (d) in Fig. 3.25, we notice two interesting observations as h_2 increases, that is, when harvesting of catfish becomes more aggressive: (1) the local maximum values of the menhaden exceed the local maximum values of the catfish; and (2) the periods for both menhaden and catfish as they exhibit their phase-shifted predator-prey behavior decrease. These observations may give rise to further inquiries on how aggressive harvesting affects the resulting predator-prey relationship between menhaden and catfish.

3.7.3 Further Numerical Investigations: Harvesting of the Blue Catfish with Equally Competitive Striped Bass

Another interesting long-term behavior of the three fish species is visualized in Fig. 3.26. Here, we consider the case when the competition effects are equal on both predators. Also, we assume that there is harvesting of the catfish. The values for h_2 are $h_2 = 0.4, 0.5, 0.8$, respectively for panels (a), (b), (c) in Figs. 3.25 and 3.26. Comparing the graphs, we observe similar time-evolution behaviors among these three species! This means that when the competition effect of the catfish is at least as strong as the striped bass $\alpha \geq \beta$, the bass loses and dies out even when harvesting of the catfish happens up to a certain point.

Panel (d) in Fig. 3.26 illustrates a very different scenario. When the competition strength between the striped bass and the catfish are equal, and when the harvesting rate $h_2 = 1.2 > 0.5 = e$ is big enough, the model shows that catfish can be driven to extinction while the menhaden and striped bass persist with a regular predator-prey relationship.

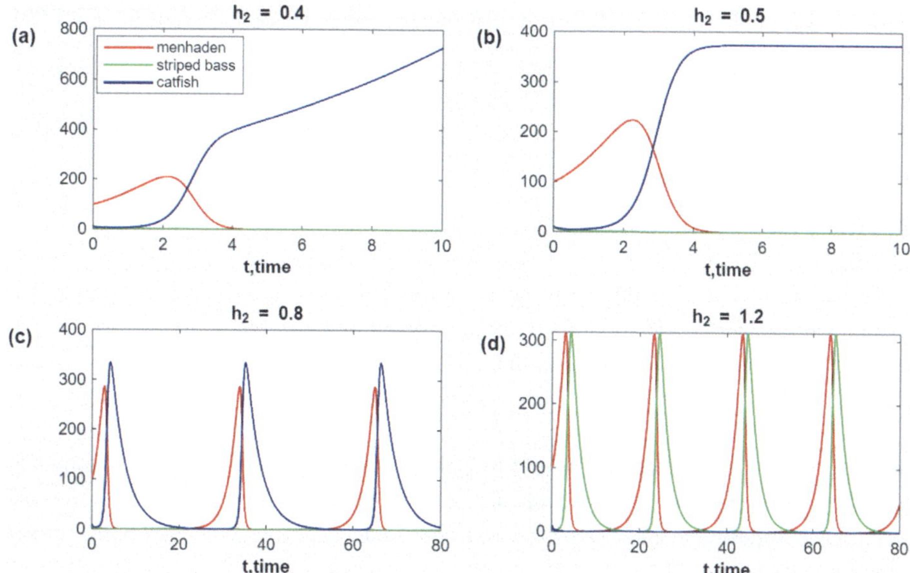

Fig. 3.26 Menhaden, striped bass, and catfish population distributions over time when a linear harvesting term $-h_2 z$ of the blue catfish is included in the model. The following values of the parameters were used: $a = d = e = 0.5$, $b = c = 0.01$, $\alpha = \beta = 0.5$ with initial conditions $(100, 10, 10)$. Panels (**a**) $h_2 = 0.4$, (**b**) $h_2 = 0.5$, (**c**) $h_2 = 0.8$, (**d**) $h_2 = 1.2$

In conclusion, this model predicts the extinction of one of the two predators depending on their competitiveness. These results are consistent with the fundamental ecological principle of competitive exclusion, also known as the Gause principle, which states that two species competing for the same limiting resources and occupying the same ecological niche cannot coexist indefinitely; eventually, one population will either be driven to extinction or adapt by occupying a different niche [48].

Exercises

3.1 Show that when the population size x is much smaller than the carrying capacity K (that is, $x \ll K$), the logistic model (3.1) reduces to the exponential growth equation $\frac{dx}{dt} = rx$. What does this tell you about early population growth?

3.2 Using Eq. (3.2) and writing the right-hand side as equation of a line $y = r - sx$, answer the following questions:

(a) When fitting the regression line to the observational *Paramecium caudatum* data in Sect. 3.2, which parameter will the y-intercept represent?
(b) How will you find the parameter s using properties of a line?
(c) What will the x-intercept represent?

Table 3.9 Observational data for *Paramecium aurelia* grown in a lab as a monoculture [6]. Population is measured in volume

Day	0	1	2	3	4	5	6	7	8	9	10
Population	1	2	3.5	5	6.2	7	7.4	7.5	7.5	7.4	7.3

3.3 (Fitting the Logistic Model)

(a) Repeat the analysis conducted in Sect. 3.2 for species *Paramecium aurelia* grown in monoculture with population distribution as given in Table 3.9.
(b) Find $RMSE$ and R^2 and interpret them. How good is the fitted model?

3.4 (Fitting the Lotka-Volterra Model to Lab Experimental Data of Two Competing Species) In this exercise you will repeat the analysis performed in Sect. 3.3.1 for two unicellular **competing** species (*Paramecium aurelia* and *Paramecium caudatum*) grown in a mixture by Gause [6].

(a) Extract the data from the *gauseR* package [4] using the command:

```
PA_PC_data =  gause_1934_science_f02_03
                       [gause_1934_science_f02_03$Treatment=="Mixture", ]
```

This will give you a data file which contains *Paramecium caudatum* data as a column named **VolumeSpecies1** and *Paramecium aurelia* data as a column named **VolumeSpeceis2**.

(b) Use the *R* program for the predator-prey system in Sect. 3.3.1 and modify it to fit the generalized Lotka-Volterra model to these new data of two competing species.

Remark Note that the optimization process is not needed in this case. Once you find the parameters using the linear regression model, you can use *ode* command (from *deSolve* package) to create the model and draw the prediction points together with the given data points. A sample program performing this fitting is provided in GitHub [8].

(c) Is there any difference in the behavior of the two set of species: predator-prey and two competing species? Explain the difference and why do you think it is like that?
(d) Using the fitted parameters, assess the long-term behavior of the two species. Does the model predicts **coexistence** or **competitive exclusion**?
(e) Find the carrying capacity and equilibrium for the dominant species or for both species if co-existence is predicted.

3.5 (Resilience of a Predator-Prey System) Find the time τ and the resilience coefficient $\mathcal{R}$ for the generalized Lotka-Volterra model, using the data from Table 3.4, also given here for convenience:

Parameter	r	a	b	e	c	d
Value	0.513	0.026	1.765	1.213	0.520	9.999

3.6 (Pearson's Correlations) Using the data from Table 3.3, calculate Pearson's sensitivity coefficients $\rho_{p_i,z}$ for each parameter p_i and each species (prey and predator).

(a) Illustrate your findings by creating figures similar to Figs. 3.11 and 3.13 for each species.
(b) Create the scatter-plots for each parameter to identify if the correlations are positive, negative, or neither.

3.7 (Spearman Correlations with Normal Sampling)

(a) Use normal distribution assumption for your sampling to calculate the Spearman correlations for each parameter when the QoI is the value of x (Atlantic menhaden) and y (striped bass) in year 2050.
(b) Verify that parameters e and d have the highest impact on prey abundance, whereas parameters r and b exhibit the greatest influence on predator abundance. To illustrate your results create figures similar to Figs. 3.15 and 3.16.

3.8 (Total-Effect Sobol' Indices and Convergence) Use the predator–prey model (3.9) from Sect. 3.4.1 with parameters $(r, a, b, e, c, d, E_1, E_2)$ and nominal values in Table 3.4. Take the QoI as either striped bass abundance y at $t_f = 50$ years or Atlantic menhaden abundance x at $t_f = 50$ years.

(a) Estimate total-effect Sobol' indices S_{T_i} for all parameters using independent *uniform* priors $p_i \sim \mathcal{U}[0.95\, p_i,\ 1.05\, p_i]$ with a baseline sample size $n = 5{,}000$ (Monte Carlo or quasi–Monte Carlo). Include uncertainty quantification (e.g., confidence intervals).
(b) Study convergence by repeating (a) for $n \in \{500, 1{,}000, 2{,}000, 5{,}000, 10{,}000\}$. State a convergence criterion and determine whether it is met for the top four parameters (by baseline S_{T_i}).

3.9 (Prior Sensitivity and Rank-Based Comparison)

(a) Replace the uniform priors with *normal* priors centered at p_i with sd $= 0.025\, p_i$, truncated to $[0.95\, p_i,\ 1.05\, p_i]$. Recompute S_{T_i} at $n = 5{,}000$ and compare rankings and uncertainties with the uniform-prior case.

(b) Using the same uniform-prior sample from Problem 1(a), compute PRCCs (with midrank transformation) and associated p-values for all parameters. Discuss agreements and discrepancies between S_{T_i} (magnitude, nonnegative) and PRCC (signed monotonic association). Identify parameters suggesting nonlinearity or interactions (e.g., large S_{T_i} but small PRCC in magnitude).

3.10 (Parameter Fixing and Variance Reduction) Using your results from Problems 3.8 and 3.9 (the baseline uniform-prior S_{T_i} at $n = 5000$ with confidence intervals, the n-sweep/convergence diagnostics, and the PRCCs with p-values), write a short memo (300–400 words) that:

(a) Recommends two parameters to fix at their nominal values to simplify calibration, justified by small S_{T_i} and narrow CIs (and, if relevant, weak PRCC or non-significant p-values).
(b) Proposes one experiment or data-collection plan to reduce output variance (e.g., narrowing the prior for a high-S_{T_i} parameter or targeting a suspected interaction), and quantifies the anticipated impact (cite your S_{T_i}, CI changes across n, and any rank shifts under normal vs. uniform priors).

3.11 (Understanding the Multi-Species Model and Its Parameters)

(a) The predator effect rates of y and z on the prey x in the model (3.26) are given by $\pm bxy$ and $\pm cxz$, respectively. How should the system be modified to consider the reality that the effect of the presence of either predators on x may not be the same as the effect of the absence of prey on the predators?
(b) If one predator is a stronger competitor than the other predator, what is the corresponding relationship between α and β?
(c) The catfish (z) is an invasive species and it will not die in the absence of the Atlantic menhaden (x). What does it mean biologically when ez replaced by $-ez$, where $e > 0$?
(d) If one assumes logistic growth for either the prey or one of the predators, how is the system (3.26) modified?
(e) Suggest how the model (3.26) could be modified so one could model various harvesting scenarios, such as constant harvesting, periodic harvesting, and delayed harvesting?

3.12 (Deriving Equilibria Solutions) Show that the equilibria solutions of the three-spices model (3.26) are given by the formulas in (3.27) and (3.28).

3.13 (Using Algebra and Calculus to Analyze Characteristic Equations) Consider the characteristic equation (3.31) and rewrite it as a cubic polynomial $g(x) = x^3 + Bx + C$, where $C < 0$. Recall from the Fundamental Theorem of Algebra that g will have at most three roots; furthermore, g can either have three real roots or one real root and a pair of complex conjugate roots.

(a) Draw possible pictures of $y = g(x)$ where g has three real roots.
(b) Draw possible pictures of $y = g(x)$ where g has one real root and a pair of complex roots.
(c) Use Intermediate Value Theorem from Calculus I to show that $y = g(x)$ has at least one positive real root.

Hint: Observe that $g(0) < 0$ and show that there is a point x_0 such that $g(x_0) > 0$. Next apply the Intermediate Value Theorem to conclude that $g(x)$ has at least one positive root.

3.14 (Harvesting of Atlantic Menhaden) What happens when the harvesting rate of Atlantic menhaden is equal to its growth rate? Analyze the system (3.32) when $h_1 = a$.

(a) Show that there are only two biologically relevant equilibria, $(0, 0, 0)$ and $\left(\frac{d}{b}, 0, 0\right)$.
(b) Show that both equilibria are unstable.
(c) Sketch a phase portrait.

3.15 (The Case When $\alpha = \beta$)

(a) Consider the three-species model (3.26) in the case when $\alpha = \beta$. Derive all equilibrium solutions and find the conditions under which they are biologically relevant.
(b) Use the linearization of the system and the stability theory to define the stability for each biologically relevant equilibrium.

3.16 (Harvesting of Catfish When $\alpha > \beta$) Consider the model when the competition interference between the striped bass and the catfish is not equal

$$\begin{cases} x' = ax - bxy - cxz \\ y' = -dy + bxy - \alpha yz \\ z' = Ez + cxz - \beta yz. \end{cases} \tag{3.38}$$

Let's assume also that the catfish is harvested at a linear rate $h_2 z$ and $\alpha > \beta > \dfrac{cd + bE}{a}$, that is, between the two competing predators, the catfish causes more harm towards the striped bass' population. Here, $E = e - h_2$. Verify the stability results about the equilibria summarized in Table 3.10. Use Theorem 3.4 Routh-Hurwitz Criteria to show the instability of the co-existence equilibrium E_4.

Table 3.10 The table presents a summary of the equilibria of model (3.38) and their stability in the case there is unequal competition interference between predators, $\alpha \neq \beta$, and when harvesting of the catfish happens $E = e - h_2$

Equilibrium point	Characteristic equation	Eigenvalues	Stability result
$Q_0(0, 0, 0)$	$(\lambda - a)(\lambda + d)(\lambda - E) = 0$	$a, -d, E$	This equilibrium point exists and is unstable for all values of E
$Q_2\left(\frac{-E}{c}, 0, \frac{a}{c}\right)$	$(\lambda + \lambda_{t+})(\lambda^2 - aE) = 0$, $\lambda_{t+} = \frac{bE + cd + \alpha a}{c}$	$\pm\sqrt{aE}, \quad -\lambda_{t+}$	$E > 0$ **not an equilibrium** $E = 0$ **unstable** $E < 0$ **unstable**
$Q_3\left(\frac{d}{b}, \frac{a}{b}, 0\right)$	$(\lambda - \lambda_{t-})(\lambda^2 + ad) = 0$, $\lambda_{t-} = \frac{bE + cd - \beta a}{b}$	$\pm i\sqrt{ad}, \quad \lambda_{t-}$	This equilibrium point exists and is unstable for all values of E
$Q_4(x^*, y^*, z^*)$ *where* $x^* = \frac{(cd + \alpha a)\beta - \alpha bE}{(\alpha + \beta)bc}$, $y^* = \frac{cd + bE + \alpha a}{(\alpha + \beta)b}$, $z^* = \frac{a\beta - (cd + bE)}{(\alpha + \beta)c}$	$\lambda^3 + B\lambda + C = 0$, where $B = b^2x^*y^* + c^2x^*z^* - \alpha\beta y^*z^*$, $C = -bc(\alpha + \beta)x^*y^*z^*$	Computing the closed-form of roots is not straightforward	This equilibrium point exists and is unstable for all values of E

List of Notations for Chap. 3

Symbol	Description	Units
t	Time	Days or years
$x(t)$	Population abundance of prey (Atlantic menhaden)	In billions
$y(t)$	Population abundance of predator (striped bass)	In billions
$z(t)$	Population abundance of invasive predator (blue catfish)	
x^*	Equilibrium of x species	
y^*	Equilibrium of y species	
z^*	Equilibrium of z spices	
r	Growth rate	
K	Carrying capacity	
a	Growth rate parameter of x	
b	Preying effect of y on x	
c	Preying effect of z on x	
d	Death rate parameter of the predator y in the absence of prey x	
e	Growth rate parameter of z	
h_1	Harvesting effect on x	
h_2	Harvesting effect on z	
α	Competition effect of z on y	
β	Competition effect of y on z	
τ	Time to return to equilibrium after perturbation	Years
$\mathcal{R}$	Resilience	Years^{-1}
λ_i	Eigenvalue	
λ_{max}	The maximum eigenvalue	
E_i	Harvesting effort	
q_i	Catchability coefficient	
J	Jacobian	
$\mathbf{Z}$	Two component vector $[x(t); y(t)]$	
$\mathbf{p}$	8-component vector of parameters, $[r; a; b; e; c; d; E_1; E_2]$	
S	Sensitivity $\mathbb{R}^{2\times 8}$ matrix	
$S_{i,j}$	Relative sensitivity $\frac{\partial x_i}{\partial p_j}\frac{p_j}{x_i}$	
$\|\|S_{ij}\|\|_2$	Sensitivity L_2 norm	
ρ	Pearson's coefficient	
ρ_s	Spearman's rank correlation coefficient	
t_f	Final integrating time	Years
$PRCC_{p_j}$	Partial rank correlation coefficient for parameter p_i	
S_{i_j}	Sobol' sensitivity indices	

List of Notations for Chap. 3 (continues)

Symbol	Description	Units
S_{T_i}	total Sobol' sensitivity index	
$RMSE$	Root mean square error	
R^2	Statistical measure of goodness of fit	
SS_{res}	The residual sum of squares $\sum_i (y_i - \hat{y}_i)^2$	
SS_{tot}	The total variability in the data around the sample mean $\sum_i (y_i - \bar{y})^2$	

References

1. Chesapeake Bay, NOAA Fisheries, https://www.fisheries.noaa.gov/topic/chesapeake-bay
2. Fisheries, Chesapeake Bay Foundation, Available at https://www.cbf.org/issues/fisheries/.
3. The MathWorks, Inc. (2024). MATLAB (Version R2024a) [Computer software]. https://www.mathworks.com.
4. Barrett, R. D. H. gauseR: Functions for fitting Lotka-Volterra models to time series data (R package version 0.1.0). (2021). https://CRAN.R-project.org/package=gauseR.
5. Mühlbauer, L.K., Schulze, M., Harpole, W.S., Clark, A.T.: gauseR: Simple methods for fitting Lotka–Volterra models describing Gause's "Struggle for Existence". *Ecology and Evolution*, 10(23), 13275–13283 (2020). https://doi.org/10.1002/ece3.6926
   Book - Open Access
6. Gause, G.F.: *The Struggle for Existence*. Williams & Wilkins, Baltimore. (1934).
7. Lehman, C., Loberg, S., Clark, A.: Quantitative Ecology: A New Unified Approach. University of Minnesota Libraries Publishing (2019). Retrieved from https://hdl.handle.net/11299/204551
8. Savatorova, V., Panayotova, I., Hallare, M. (2025). Mathematical-modeling-in-Life-Sciences-A-Practical-Project-Based-Approach [Source code]. GitHub. https://github.com/ViktoriaLS/Mathematical-modeling-in-Life-Sciences-A-Practical-Project-Based-Approach
9. Kingsland, S.E.: Modeling Nature: Episodes in the History of Population Ecology, 2nd edn. University of Chicago Press, Chicago. (1995).
10. Lotka, A.J.: Elements of Physical Biology. Williams & Wilkins, Baltimore (1925)
11. Volterra, V., Fluctuations in the abundance of a species considered mathematically. *Nature*, 118(2972), 558–560. (1926).
12. R Core Team. R: A language and environment for statistical computing. R Foundation for Statistical Computing. (2024). https://www.R-project.org/
13. RStudio Team. RStudio: Integrated Development Environment for R. Posit Software, PBC. (2023). https://posit.co/
14. Atlantic States Marine Fisheries Commission, *Status of the Stocks Overview*, Arlington, VA, (2025). Available at: https://asmfc.org/resources/stock-assessment/status-of-the-stocks-overview/, Accessed 2025-09-23.
15. Chesapeake Bay Trust and University of Maryland Center for Environmental Science. Forage indicators and consumption profiles for dominant prey species in Chesapeake Bay (Report No. FY14 Forage Fish Indicator Metric Development). Chesapeake Bay Trust. (2016). Available at: https://cbtrust.org/wp-content/uploads/FY14-Forage-fish-indicator-metric-development-1.pdf

16. Pimm, S.L. and J.H. Lawton, *On feeding on more than one trophic level*, Nature, 275: 542–544, (1978).
17. Saltelli, A., Chan, K., Scott, EM. *Sensitivity analysis*, Wiley series in probability and statistics. Wiley, Chichester; New York. (2000)
18. Saltelli, A., Ratto, M., Andres, T., Campolongo, F., Cariboni, J., Gatelli, D., Saisana, J., Tarantola, S. *Global Sensitivity analysis: The primer*, Wiley, LTD., Chichester. (2008)
19. Razavi, S., Jakeman, A., Saltelli, A., Prieur,C., Iooss, B., Borgonovo, E., Plischke, E., Lo Piano, S., Iwanaga, T., Becker, W., Tarantola, S., Guillaume, J.H.A. , Jakeman, J., Gupta, H., Melillo, N., Rabitti, G., Chabridon, V., Duan, Q., Sun, X., Smith, S., Sheikholeslami, R., Hosseini, N., Asadzadeh,M., Arnald, P., Kucherenko, S., Maier, H.R., The Future of Sensitivity Analysis: An essential discipline for systems modeling and policy support. *Environmental Modelling and Software* 137: 104954. (2021).
20. Savatorova, V., Exploring parameter sensitivity analysis in mathematical modeling with ordinary differential equations. *CODEE Journal* **16**, Article 4 (2023)
21. Panayotova I.N., Herrmann, J., Kolling, N., Bioeconomic analysis of harvesting within a predator-prey system: A case study in the Chesapeake Bay fisheries. *Ecological Modeling*, 480, 110330. (2023).
22. Drew, K., Cieri, M., Schueller, A. M., Buchheister, A., Chagaris, D., Nesslage, G., McNamee, J. E., Uphoff, J. H., et al., *Balancing Model Complexity, Data Requirements, and Management Objectives in Developing Ecological Reference Points for Atlantic Menhaden*. Frontiers in Marine Science, 8:608059. (2021). https://doi.org/10.3389/fmars.2021.608059
23. McKay, M.D., Beckman, R.J., Conover, W.J., Comparison of 3 Methods for Selecting Values of Input Variables in the Analysis of Output from a Computer Code. *Technometrics* 21(2): 239–245. (1979).
24. Larsen, R.J., Marx, M.L., *An Introduction to Mathematical Statistics and Its Applications*, Pearson. (2018).
25. James, G., Witten, D., Hastie, T. and Tibshirani, R., *An Introduction to Statistical Learning with Applications in R*, Springer, Berlin. (2017).
26. Kendall, M.G., Partial rank correlation. *Biometrica*, 32(3/4): 277–283. (1942).
27. Sobol', I.M., Sensitivity estimates for nonlinear mathematical models. *Math. Model. Comput.Exp.*, 1(4):407–414. (1993).
28. Saltelli, A., Annoni, P., Azzini, I., Campolongo, F., Ratto, M. and Tarantola, S., Variance based sensitivity analysis of model output. Design and estimator for the total sensitivity index. *Computer physics communications*, 181 (2):259–270. (2010).
29. Fisher, R. A., & Mackenzie, W.A., Studies in crop variation. ii.the manurial response of different potato varieties. *The Journal of Agricultural Science*, 13(3): 311–320. (1923).
30. Sobol, I.M., Global sensitivity indices for nonlinear mathematical models and their Monte Carlo estimates. *Mathematics and Computers in Simulation*, 55, p. 271–280. (2001).
31. Pianosi, F., Sarrazin, F., Wagener, T., A Matlab toolbox for Global Sensitivity Analysis. *Environmental Modeling & Software*, 70: 80–85. (2015).
32. Herman, J. and Usher, W. SALib: An open-source Python library for sensitivity analysis. *Journal of Open Source Software*, 2(9), 97. (2017). doi: 10.21105/joss.00097
33. Jolliffe,I.T. (2002) *Principal component analysis for special types of data*, Springer, New York.
34. Chesapeake Bay Foundation, *Atlantic Menhaden*, https://www.cbf.org/about-the-bay/more-than-just-the-bay/chesapeake-wildlife/menhaden/
35. Striped Bass Conservation Regulations Set for Spring 2020, PropTalk, August, (2020). Available at https://www.proptalk.com/striped-bass-conservation-regulations-set-spring-2020
36. Schloesser, R. W., Fabrizio, M. C., Latour, R. J., Garman, G. C., Greenlee, B., Groves, M., Gartland, J. (2011). Ecological role of blue catfish in Chesapeake Bay tributaries and potential effects on native species. Transactions of the American Fisheries Society, 140(5), 1391–1403.

37. Fabrizio, M. C., Tuckey, T. D., Norris, A. J., Latour, R. J. (2018). Trophic ecology of invasive blue catfish in Chesapeake Bay. Marine and Coastal Fisheries, 10(1), 90–108.
38. NOAA Fisheries News. March (2018). Available at https://www.fisheries.noaa.gov/feature-story/bay-invaders-blue-catfish-fishery.
39. Panayotova, I. N., Hallare, M., Modeling the Ecological Dynamics of a Three-Species Fish Population in the Chesapeake Bay. CODEE Journal **14**, Article 2 (2021). https://doi.org/10.5642/codee.202114.01.02
40. Arnold, V.I. *Ordinary Differential Equations*, Springer-Verlag, 1992.
41. Guckenheimer, J. and Holmes, P. *Nonlinear Oscillations, Dynamical Systems, and Bifurcations of Vector Fields*, Springer, (1983).
42. Perko, L. *Differential Equations and Dynamical Systems*, Springer-Verlag, New York, (2001).
43. Willems, J. L. *Stability Theory of Dynamical Systems*, Wiley, New York, (1970).
44. Atlantic menhaden, *Atlantic States Marine Fisheries Commission.* Retrieved November 15, 2021, from http://www.asmfc.org/species/atlantic-menhaden
45. National Oceanic and Atmospheric Administration, *Blue catfish: Invasive and Delicious*, (n.d.) Retrieved November 10, (2021) Available at https://www.fisheries.noaa.gov/feature-story/blue-catfish-invasive-and-delicious
46. ASMFC Approves Resolution on Non-native Invasive Catfish, *ASMFC Fisheries Focus*, (2011). Available at http://www.asmfc.org/uploads/file/septOct2011.pdf.
47. Orth, D., Y. Jiao, J. Schmitt, C.Schmitt, *Dynamics and Role of Non-native Blue Catfish Ictalurus furcatus in Virginia's Tidal Rivers Final Report*, (2017). https://doi.org/10.13140/RG.2.2.35917.54246
48. Hardin, G. The Competitive Exclusion Principle, *Science, New Series*, 131, 3409, pp. 1292–1297, (1960).

Modeling Cancer Growth and Treatment 4

4.1 Introduction

Cancer can be viewed as a problem of interacting populations—tumor cells, immune cells, and therapeutic agents. Thinking in terms of populations makes ordinary differential equations (ODEs) a natural framework for modeling these interactions. Each term in an ODE model corresponds to a specific biological process, making the underlying assumptions transparent. Tumor and immune populations influence one another through growth, decay, and feedback terms, and classical tools such as equilibria, stability, and nullclines help us understand how parameter values shape the outcomes. Numerical simulations in MATLAB complement the analysis and make these models accessible in the classroom.

Mathematical oncology has developed over several decades, beginning with early tumor growth laws in the 1950s–1960s and continuing with tumor–immune interaction models in the 1990s and 2000s. A key contribution was the work of Kuznetsov et al. [1], who introduced nonlinear immune response functions and bistable tumor dynamics. Their framework forms the basis of many modern immunotherapy models.

More recently, mathematical oncology has emerged as a field that connects theoretical modeling with clinical questions. As described in the *2019 Mathematical Oncology Roadmap* [2], a central motivation is to use mechanistic models—such as the ODE systems in this chapter—to help interpret patient-specific tumor dynamics and treatment responses. The roadmap also highlights the growing integration of mechanistic models with data-driven methods so that treatment strategies can adapt as the tumor evolves.

Recent work has also emphasized the importance of evolutionary ideas in cancer therapy. Belkhir et al. [3] review how eco-evolutionary principles can guide adaptive treatments designed to delay or prevent drug resistance. In their framework, tumors behave as evolving ecosystems, and Lotka–Volterra competition models—closely related to those used here—help track the balance between drug-sensitive and drug-resistant

I. Panayotova et al., *Mathematical Modeling in Life Sciences*, Synthesis Lectures on Mathematics & Statistics, https://doi.org/10.1007/978-3-032-17486-4_4

subpopulations. Under continuous therapy, sensitive cells may be eliminated first, allowing resistant cells to expand, which often leads to recurrence. Alternating treatment and rest cycles can maintain a sufficient population of drug-sensitive cells to suppress resistant ones, a "containment strategy" aimed at stabilizing tumor burden rather than eliminating it outright. This approach aligns with a broader goal of mathematical oncology: use modeling and simulation to understand tumor dynamics and inform treatment strategies that respond to evolutionary pressures.

This chapter begins with classical tumor growth laws (Sect. 4.2), then develops interaction models based on tumor–immune dynamics (Sects. 4.3–4.4), and concludes with a melanoma immunotherapy case study (Sect. 4.5). Numerical simulations illustrate the dynamics and support the analytical findings (Sect. 4.6). Finally, we relate the mathematical terms to the *Hallmarks of Cancer* (Sect. 4.7), illustrating how the underlying biology is reflected in the equations.

The models in this chapter do not incorporate pharmacokinetics or pharmacodynamics (PK/PD); in particular, we do not model how drug concentrations change over time or how dosing schedules modify treatment effects.

4.2 Cancer Growth Models: Practical Considerations

A natural first step in modeling tumor–immune interactions is to examine how tumor cells grow in isolation, without immune response or treatment effects. This baseline description can then be expanded to incorporate additional biological processes as needed.

There is no single universal law of tumor growth, but several classical models capture features observed in experimental and clinical settings: *exponential*, *logistic*, *Gompertz*, and *von Bertalanffy* [4]. Each model reflects different assumptions about cell proliferation, resource limitations, and growth inhibition. Table 4.1 summarizes these growth laws and their analytical solutions. Exercises at the end of this chapter help students compare these models more closely.

Model selection depends on tumor type, size, and growth phase. Early-stage tumors often grow nearly exponentially, so an exponential form describes that phase well. As

Table 4.1 Classical tumor growth laws with closed-form solutions

Growth law	ODE	Solution
Exponential	$\dfrac{dC}{dt} = a\,C$	$C(t) = C_0\,e^{at}$
Logistic	$\dfrac{dC}{dt} = a\,C\,(1 - bC)$	$C(t) = \dfrac{C_0\,e^{at}}{1 + bC_0\,(e^{at} - 1)}$
Gompertz	$\dfrac{dC}{dt} = a\,C\,\ln\big(\frac{1}{bC}\big)$	$C(t) = \dfrac{1}{b}\,\exp\big(\ln(bC_0)\,e^{-at}\big)$
von Bertalanffy	$\dfrac{dC}{dt} = a\,C\big(1 - (bC)^r\big)$	$C(t) = \dfrac{1}{b}\left[1 + \dfrac{1-(bC_0)^r}{(bC_0)^r}\,e^{-art}\right]^{-1/r}$

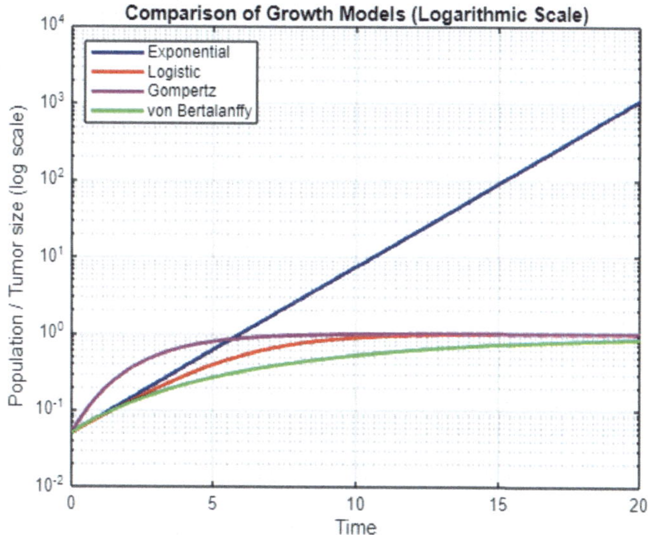

Fig. 4.1 Comparison of tumor growth models on a logarithmic scale. The exponential model appears as a straight line (in blue) because its logarithm increases linearly with time. The logistic, Gompertz, and von Bertalanffy models exhibit sigmoidal (S-shaped) behavior due to self-limiting growth; this reflects saturation as the population approaches the carrying capacity K

tumors enlarge, nutrient limitations and cell crowding slow their expansion. The logistic and Gompertz models capture this self-limiting behavior, while the von Bertalanffy model introduces an additional parameter that provides extra flexibility. See Fig. 4.1 for a visualization of these four models on a single graph. For example, the exponential model has the form

$$C(t) = C_0 e^{at},$$

where a is the growth rate and C_0 is the initial population. Taking the logarithm gives

$$\log C(t) = \log(C_0 e^{at}) = \log C_0 + at,$$

which is linear in time t.

One useful feature of these classical growth laws is that they have *closed-form solutions*, so the tumor size $C(t)$ can be written explicitly as a function of time. When data are available, model parameters can be estimated by fitting tumor measurements to the analytic expression. A simple workflow for model fitting is:

1. Convert tumor volume or diameter data into approximate cell counts.
2. Select a candidate growth model from Table 4.1.
3. Use the model's explicit solution to evaluate $C(t)$ at measurement times.

4. Adjust parameters to minimize the difference between model predictions and data.

This process yields parameter estimates that characterize intrinsic tumor properties such as the growth rate and carrying capacity. Two models may fit data equally well—for instance, logistic and von Bertalanffy—but the logistic model is often preferred because it achieves similar accuracy with fewer parameters. When choosing among competing models, the general advice is to balance *biological realism* against *mathematical simplicity*. In accordance with Occam's principle, the more parsimonious model—the one requiring fewer parameters or assumptions—is typically favored when explanatory power is comparable.

These classical models require only Calculus I/II tools. They illustrate how biological assumptions shape equations and how mathematical forms connect to data. Although this chapter does not carry out parameter estimation in detail, students can use these models as templates for exploratory assignments or undergraduate research. See the previous chapter for a discussion of data fitting and parameter analysis.

Once intrinsic parameters have been estimated, they form the foundation for more realistic models. External influences—such as immune response, treatment, or angiogenesis—can then be introduced systematically, one term at a time.

A typical modeling process may proceed as follows:

1. Fit tumor growth data to a classical law to estimate intrinsic parameters.
2. Interpret these parameters biologically (e.g., growth rate, carrying capacity).
3. Add immune interaction terms to represent tumor–immune dynamics (see Sect. 4.3).
4. Introduce treatment effects such as chemotherapy, radiotherapy, or surgery (see Sect. 4.4.3).
5. Optionally add other effects such as resource competition or angiogenesis.
6. Analyze the resulting system using ODE techniques (equilibria, stability, and long-term behavior, discussed in Sect. 4.5).
7. Simulate the system numerically (e.g., with MATLAB); sample codes are available in the chapter repository [5]. See Sect. 4.6.
8. Interpret and compare outcomes across parameter regimes.

Later in the chapter, Sect. 4.7, we connect mathematical terms in the model to the Hallmarks of Cancer [6] to demonstrate how the biological mechanisms are represented within the equations.

4.3 Motivation for Tumor-Immune Interaction Models

In the previous section, we reviewed classical models of tumor growth in isolation. For much of the twentieth century, cancer research and treatment followed a similar perspective: the tumor was viewed as a self-contained system. The dominant treatment strategies—surgery, radiation therapy, and chemotherapy—focused on *removing* or *destroying* tumors directly, with little consideration for the immune system's role [7]. Surgery physically reduced tumor burden, radiation damaged cellular DNA, and chemotherapy used cytotoxic drugs to halt cell division. These methods were grounded in the belief that tumor behavior and treatment response depended primarily on the tumor itself.

This reductionist view was reflected in early mathematical models. Tumors were represented as single populations governed by logistic or Gompertz growth, with treatment effects modeled as external "kill" terms. Because the immune system was not yet recognized as a major factor in cancer progression, it was omitted from both therapeutic design and mathematical formulation.

A turning point came with the *immune surveillance hypothesis*, which proposed that the immune system continually monitors tissues and eliminates abnormal cells before they become malignant [7]. Experimental evidence soon confirmed that while some tumors are successfully eliminated, others evade or suppress immune activity. This insight fundamentally changed how cancer was understood: tumor growth depends not only on cancer cell proliferation but also on the dynamic interplay between tumor and immune populations.

This realization paved the way for the development of *immunotherapy*, a class of treatments designed not merely to kill tumor cells but to restore or enhance immune function. De Pillis and Radunskaya [4] classify immunotherapies into three broad categories:

(1) **Immune response modifiers**, which directly stimulate or suppress immune activity.
(2) **Monoclonal antibodies**, which help the immune system recognize and target specific tumor antigens.
(3) **Cancer vaccines**, which train immune cells to identify and attack tumor-associated antigens.

The models in [4] focus primarily on Categories (1) and (3), introducing immune stimulants and vaccine terms to enhance immune recruitment. Although monoclonal antibodies are noted as promising, they are not explicitly modeled in their framework. By contrast, our melanoma case study in Sect. 4.5, based on the work of Pennisi [8], centers on Category (2) therapies—specifically, anti-CD137 monoclonal antibodies that amplify cytotoxic T-cell activity against tumor cells.

In the next section, we introduce a sequence of models that we call *Base Tumor-Immune Models*. We begin with **Base Tumor-Immune Model 1**, the simplest predator—prey formulation, which assumes exponential tumor growth and direct immune killing. Although biologically limited, this minimal model provides a clear setting to review classical ODE tools such as nullclines, linearization, and phase plane analysis. **Base Tumor-Immune Model 2** incorporates logistic tumor growth and an immune influx term, reflecting biological constraints on tumor expansion and a more realistic immune response. This model serves as the foundation for adding treatment terms representing surgery, radiation, and chemotherapy, see Sect. 4.4.3.

Finally, **Base Tumor-Immune Model 3** presents the classic Kuznetsov model of tumor immunotherapy [1]. This system introduces nonlinear immune stimulation and captures richer dynamics, for example, immune thresholds, tumor dormancy, and immune escape. Although it represents the most biologically detailed system in this sequence, our treatment models in Sect. 4.4.3 build instead on **Base Tumor-Immune Model 2** for simplicity. Students may explore the Kuznetsov model as a Base Tumor-Immune Model and apply ideas presented here.

4.4 Predator–Prey in Cancer Treatments

A straightforward way to describe cancer treatment mathematically is to add a loss term to the tumor population, similar to predator–prey models where predators directly remove prey.

In this section, we develop a stepwise sequence of Base Tumor-Immune Models:

- **Base Tumor-Immune Model 1: Minimal Predator–Prey.** A Lotka–Volterra framework with exponential tumor growth and immune killing. This minimal model is used to demonstrate fundamental ODE tools such as nullclines, linearization, and phase plane analysis.
- **Base Tumor-Immune Model 2: Logistic Growth + Immune Influx.** This model improves on Base model 1 by incorporating tumor carrying capacity and baseline immune influx, making it more biologically realistic while still easy to analyze. It serves as our *main platform* for introducing treatment effects such as surgery, radiation, chemotherapy, and combo-therapy.
- **Base Tumor-Immune Model 3: Kuznetsov Model (Classic).** This well-known model [1] introduces nonlinear immune stimulation and saturation, leading to richer tumor–immune dynamics (e.g., thresholds, dormancy, escape). Although not the main focus here, it is presented in the final subsection as an alternative base model that students can explore.

4.4.1 Base Tumor-Immune Model 1: Minimal Predator–Prey

Immunotherapy works by stimulating the patient's immune system to recognize and attack tumor cells. Immune *effector* cells are produced in the bone marrow and circulate even in the absence of a specific threat. They can recognize abnormal cells, including tumor cells, and kill them.

Let $C(t)$ represent tumor cells (prey) and $E(t)$ represent effector cells (predator). Assume that tumor cells grow exponentially, while encounters between tumor and effector cells remove tumor cells at a rate proportional to pEC. Effector cells are stimulated by tumor presence and decay naturally at rate d. The modeling system is

$$\begin{cases} \dfrac{dC}{dt} = rC - pEC, \\ \dfrac{dE}{dt} = sEC - dE, \end{cases} \qquad r, p, s, d > 0. \tag{4.1}$$

This is the classical Lotka–Volterra predator–prey system, which we use as a first model for tumor–immune interactions. Its structure makes it easy to apply phase-plane methods, examine equilibria, and explore stability properties.

Nullclines and Equilibria: The nullclines are simple horizontal and vertical lines:

- $C = 0$ or $E = r/p$, from $\dfrac{dC}{dt} = 0$.
- $E = 0$ or $C = d/s$, from $\dfrac{dE}{dt} = 0$.

Their intersection gives two equilibria:

$$(C^*, E^*) = (0, 0) \quad \text{and} \quad (C^*, E^*) = \left(\frac{d}{s}, \frac{r}{p}\right).$$

Linearization and Phase Plane: The Jacobian matrix is

$$J(C, E) = \begin{pmatrix} r - pE & -pC \\ sE & sC - d \end{pmatrix}.$$

By plugging-in the equilibrium states into the Jacobian and analyzing the eigenvalues of the resulting matrices, the origin is a saddle point (one positive and one negative eigenvalue), while the interior equilibrium $\left(\frac{d}{s}, \frac{r}{p}\right)$ requires some further handling because its eigenvalues are $\pm i\sqrt{rd}$. Using standard arguments in ODE or dynamical systems (see for example, [9] or [10]), this interior equilibrium is a center with closed orbits and neutral stability. Figure 4.2 illustrates these dynamics via a phase plane produced using MATLAB's `ode45`.

Because there is no stable tumor-free equilibrium, the tumor and effector populations coexist at the interior equilibrium. This motivates moving to a more realistic model structure.

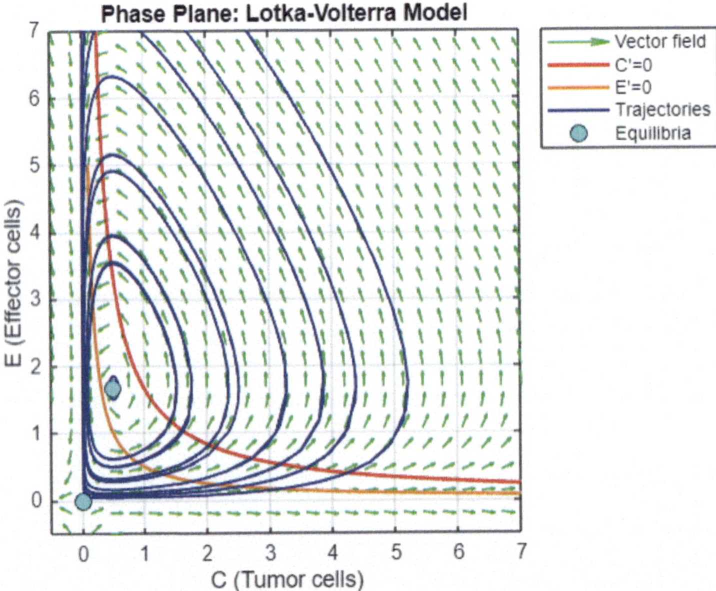

Fig. 4.2 Phase plane of Base Tumor-Immune Model 1 using the parameter values given in Table 4.4 in the Appendix. The origin represents the tumor-free equilibrium and acts as a saddle point, while the interior point represents tumor-persistent equilibrium and is a center. Trajectories circulate around the interior equilibrium, indicating neutral coexistence between tumor and effector cell populations which is a behavior characteristic of simple predator–prey systems

4.4.2 Base Tumor-Immune Model 2: Logistic Growth and Immune Influx

To improve biological realism, we add two key assumptions:

- *Logistic tumor growth* with carrying capacity K to reflect resource limitation.
- *Baseline immune influx* s and stimulation cC to represent constant and tumor-dependent immune response.

The system becomes

$$\begin{cases} \dfrac{dC}{dt} = rC\left(1 - \dfrac{C}{K}\right) - pEC, \\ \dfrac{dE}{dt} = s + cC - dE. \end{cases} \tag{4.2}$$

This model preserves analytical tractability. We will use this baseline model when we incorporate time-dependent cancer treatment interventions. In Fig. 4.3, MATLAB simulations using `ode45` show how Base Tumor–Immune Model 2 behaves without treatment. With weak immune response (left), the tumor settles to a positive level. With strong

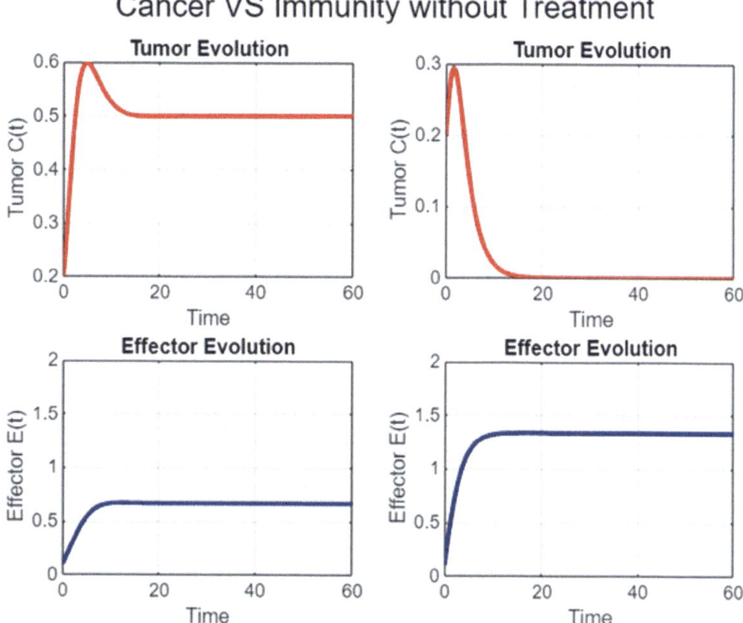

Fig. 4.3 Tumor–immune dynamics for Base Tumor-Immune Model 2 using parameter values given in Table 4.5 in the Appendix. Here, no treatments have been incorporated yet. Left: **weak** immune response leads to coexistence and recurrence. Right: **strong** immune response results in tumor elimination

immune response (right), the effector cells rise enough to clear the tumor. The two cases illustrate how immune strength alone can decide the outcome.

For clarity, cancer treatments will be incorporated by modifying Base Tumor-Immune Model 2:

$$\begin{cases} \dfrac{dC}{dt} = \underbrace{rC\left(1 - \dfrac{C}{K}\right)}_{\text{tumor proliferation}} - \underbrace{pEC}_{\text{immune-mediated killing}} - \text{cancer treatment}, \\ \dfrac{dE}{dt} = \underbrace{s}_{\text{baseline influx}} + \underbrace{cC}_{\text{tumor stimulation}} - \underbrace{dE}_{\text{effector decay}} . \end{cases} \tag{4.3}$$

4.4.3 Cancer Treatments on Base Tumor-Immune Model 2

Next we model the three classical cancer treatments:

- **Surgery**: instantaneous tumor reduction at $t = t_s$.
- **Radiation therapy**: tumor death over a treatment window.

- **Chemotherapy**: death of both tumor and immune populations over a treatment window.
- **Combination therapy**: a mixture of the above mechanisms.

We now incorporate treatments one at a time, adding only the terms that represent their direct biological effects on the tumor and immune cells. We start with the simplest case: surgery.

Example 4.1 (Surgery-Immune Model) Surgery removes a fixed fraction σ of the tumor mass at a specific time $t = t_s$. This is modeled as an instantaneous pulse via the Dirac delta distribution:

$$\begin{cases} \dfrac{dC}{dt} = r\,C\left(1 - \frac{C}{K}\right) - p\,E\,C\, - \sigma\,C\,\delta(t - t_s), \\ \dfrac{dE}{dt} = s + c\,C - d\,E. \end{cases} \tag{4.4}$$

The surgery is modeled by the term $-\sigma C\,\delta(t - t_s)$ in the tumor equation. To see how this affects the solution at $t = t_s$, we integrate across a small interval $[t_s - \varepsilon,\ t_s + \varepsilon]$ around the surgery time:

$$\int_{t_s-\varepsilon}^{t_s+\varepsilon} \frac{dC}{dt}\,dt = \int_{t_s-\varepsilon}^{t_s+\varepsilon} \left[r\,C\left(1 - \frac{C}{K}\right) - p\,E\,C - \sigma\,C\,\delta(t - t_s)\right] dt.$$

The left-hand side simplifies to

$$\int_{t_s-\varepsilon}^{t_s+\varepsilon} \frac{dC}{dt}\,dt = C(t_s^+) - C(t_s^-),$$

which is the jump in $C(t)$ across t_s. Now, for the right-hand side, as $\varepsilon \to 0$, the logistic and immune killing terms contribute negligibly since they are finite. Only the δ term remains:

$$-\sigma \int_{t_s-\varepsilon}^{t_s+\varepsilon} C(t)\,\delta(t - t_s)\,dt = -\sigma\,C(t_s^-).$$

Combining, we have $C(t_s^+) - C(t_s^-) = -\sigma\,C(t_s^-)$, and therefore (see [11])

$$C(t_s^+) = (1 - \sigma)\,C(t_s^-).$$

The Dirac delta function is a convenient way to model an instantaneous intervention like surgery as a cancer treatment. It makes the jump at t_s mathematically precise without altering the dynamics at other times. (Note that some students may not have previously encountered the Heaviside function or the Dirac delta distribution—which is not a function in the usual sense. A useful math-modeling-first textbook that introduces these objects through IVPs with discontinuous and impulsive forcing, solved via the Laplace transform, is [12].).

However, in MATLAB, we do not include the Dirac delta term $\delta(t - t_s)$ explicitly. Since δ is a distribution rather than an ordinary function, it is not directly compatible with numerical solvers such as `ode45`. In MATLAB, `dirac` is symbolic, not numeric. Instead, we implement its effect exactly by integrating the system up to the surgery time t_s and then applying the jump condition

$$C(t_s^+) = (1 - \sigma)\, C(t_s^-), \qquad E(t_s^+) = E(t_s^-),$$

where σ represents the fraction of tumor removed during surgery. This approach is both efficient and numerically stable.

Equivalently, the system can be described piecewise in time as

$$\begin{cases} \dfrac{dC}{dt} = r\,C\left(1 - \frac{C}{K}\right) - p\,E\,C, \\ \dfrac{dE}{dt} = s + c\,C - d\,E, \end{cases} \qquad t \neq t_s,$$

with the above jump condition. The tumor population $C(t)$ drops instantaneously by a fraction σ. Since the effector equation has no δ term, $E(t)$ remains continuous, that is, $E(t_s^+) = E(t_s^-)$.

In Fig. 4.4, we use MATLAB's `ode45` to simulate (4.4). Surgery is mathematically modeled as an instantaneous event happening at $t = 20$. (Clinical experience, of course, tells us that surgery is not an instantaneous event!) The surgery itself doesn't determine whether the tumor comes back or not. Instead, what matters is what happens after the tumor is reduced via surgery. This is where we use the mathematics rationale (and resulting MATLAB code) to explain the cancer treatment clinically: the surgery event *acts like a reset* and a different initial-value problem (IVP) governs the system. Surgery pushes the system to a new starting point with a smaller tumor. Moreover, we observe from the evolution plots that prior to surgery (4.4), the red cancer curves in the top row seems to have a stable steady-state with positive C. Because of the cancer-treatment intervention, the system could have either tumor recurrence (positive C stable level on the upper left) or tumor elimination (zero C stable level on the upper right). If the immune response is *weak* (lower blue effector levels), the tumor grows back to a positive cancer-level. But if the immune response is *strong* (higher blue effector levels), it can clear the remaining tumor completely. In other words, the immune system decides the final outcome after surgery happens.

Example 4.2 (Radiation Therapy-Immune Model) Radiation therapy is applied over a fixed interval $[t_r,\ t_r + T_r]$ and increases tumor cell death during this period. To represent

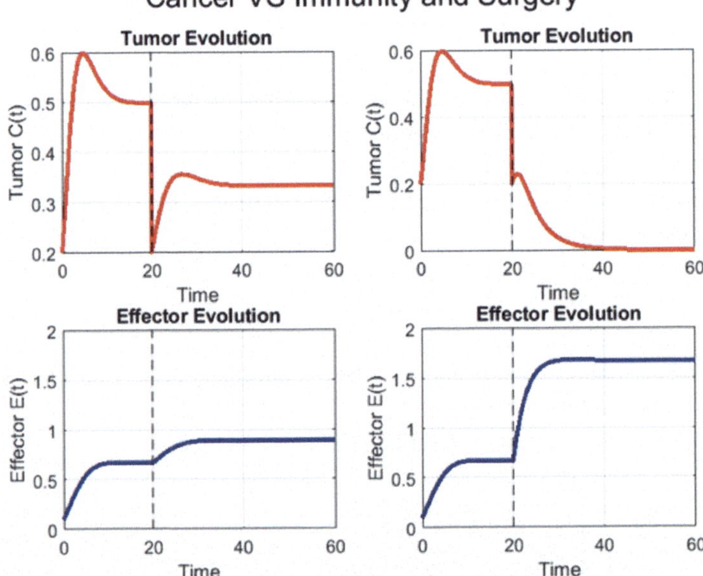

Fig. 4.4 Surgery and immune response outcomes. Surgery is performed at time $t = 20$, causing an instantaneous reduction in the tumor population. The top row depicts the evolution of the tumor population $C(t)$, and the bottom row shows the effector population $E(t)$. The **left panels** show tumor recurrence, while the **right panels** show tumor elimination after surgery. When the immune response is *weak*, surgery removes a fraction of the tumor, but the remaining population regrows over time, approaching a positive steady state. When the immune response is *strong*, the residual tumor is cleared, and the system converges to a tumor-free steady state. Parameter values used to create this figure are given in Table 4.5 in the Appendix

this (see [11]), we define an indicator function

$$u(t) = \begin{cases} 1, & t_r \le t \le t_r + T_r, \\ 0, & \text{otherwise.} \end{cases}$$

The parameter ρ denotes the per capita tumor death rate induced by radiation. The model becomes

$$\begin{cases} \dfrac{dC}{dt} = r\,C\left(1 - \frac{C}{K}\right) - p\,E\,C - \rho\,C\,u(t), \\ \dfrac{dE}{dt} = s + c\,C - d\,E. \end{cases} \tag{4.5}$$

During the treatment window $[t_r, t_r + T_r]$, the additional term $-\rho C\,u(t)$ is active and increases tumor death. Once the radiation stops, $u(t)$ returns to zero and the system reverts to the base predator–prey model. The only additional parameters introduced here are ρ (radiation-induced death rate), t_r (start time), and T_r (duration of treatment).

Since (4.5) is non-autonomous as written, we can equivalently write it as three autonomous initial-value problems (IVPs): **before**, **during**, and **after radiation**. Each system has constant right-hand side terms and is solved over a specific time interval.

IVP 1: Before radiation $(0 \le t < t_r)$

$$\begin{cases} \frac{dC}{dt} = r\,C\left(1-\frac{C}{K}\right) - p\,E\,C, \\ \frac{dE}{dt} = s + c\,C - d\,E, \end{cases} \quad \text{with initial condition } C(0) = C_0,\ E(0) = E_0.$$

IVP 2: During radiation $(t_r \le t \le t_r + T_r)$

$$\begin{cases} \frac{dC}{dt} = r\,C\left(1-\frac{C}{K}\right) - p\,E\,C - \rho\,C, \\ \frac{dE}{dt} = s + c\,C - d\,E, \end{cases} \quad \text{with initial condition } C(t_r) = C_1,\ E(t_r) = E_1,$$

where (C_1, E_1) are obtained from IVP 1 at $t = t_r$.

IVP 3: After radiation $(t > t_r + T_r)$

$$\begin{cases} \frac{dC}{dt} = r\,C\left(1-\frac{C}{K}\right) - p\,E\,C, \\ \frac{dE}{dt} = s + c\,C - d\,E, \end{cases} \quad \text{with initial condition } C(t_r+T_r) = C_2,\ E(t_r+T_r) = E_2,$$

where (C_2, E_2) are obtained from IVP 2 at the end of the radiation window.

In Fig. 4.5, we see evolution plots produced by MATLAB simulation for Eqs. (4.5). Observe that for the pictures on the right panels, the tumor population has a transient bump before eventually decaying as the effector cell population ramps up. Also, comparing radiotherapy with surgery, we see that radiotherapy works over time interval, slowly shrinking the tumor during which time, the the immune response has a time to react and do its job of suppressing the tumor. This means radiotherapy not only reduces the tumor but can also shape how the immune system reacts in killing the cancer. Surgery, on the other hand, makes one sudden change and then relies on the immune system to handle what's left. See [13] to learn more about radiotherapy impact on tumor-immune interactions.

Example 4.3 (Chemotherapy-Immune Model) Chemotherapy in our models is represented by direct kill terms on tumor *and* immune cells. It is applied over a fixed interval $[t_c, t_c + T_c]$. As before, define an indicator function

$$u(t) = \begin{cases} 1, & t_c \le t \le t_c + T_c, \\ 0, & \text{otherwise.} \end{cases}$$

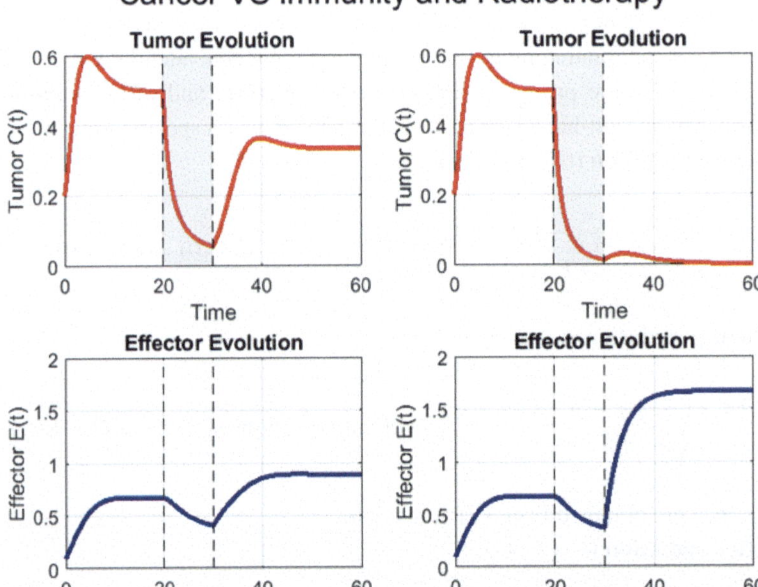

Fig. 4.5 **Radiation therapy and immune response outcomes.** Radiation therapy is administered during the time interval [20, 30]. The top row depicts the evolution of the tumor population $C(t)$, and the bottom row shows the effector population $E(t)$. The **left panels** show tumor recurrence, while the **right panels** show tumor elimination after the radiation treatment. When the immune response is *weak*, radiation reduces the tumor size during the treatment window $[t_r, t_r + T_r]$, but the tumor eventually regrows, approaching a positive steady state. When the immune response is *strong*, the remaining tumor is cleared, and the system converges to a tumor-free steady state. Parameter values used to create this figure are given in Table 4.5 in the Appendix

Let γ_C and γ_E be the per capita death rates of tumor and effector cells, respectively, due to chemotherapy. The model becomes

$$\begin{cases} \dfrac{dC}{dt} = r\,C\left(1 - \frac{C}{K}\right) - p\,E\,C - \gamma_C\,C\,u(t), \\ \dfrac{dE}{dt} = s + c\,C - d\,E - \gamma_E\,E\,u(t). \end{cases} \tag{4.6}$$

During the treatment window, the additional terms $-\gamma_C Cu(t)$ and $-\gamma_E Eu(t)$ are active and represent chemotherapy-induced death in both populations. Once chemotherapy stops, $u(t)$ returns to zero and the system again follows the base predator–prey dynamics. Effector cells recover through the baseline influx s and stimulation cC.

In Fig. 4.6, we see evolution plots produced by MATLAB simulation (via ode45) for Eqs. (4.6). Similar to radiotherapy, chemotherapy acts over a time interval, continuously reducing the tumor population during treatment. However, unlike radiotherapy, chemother-

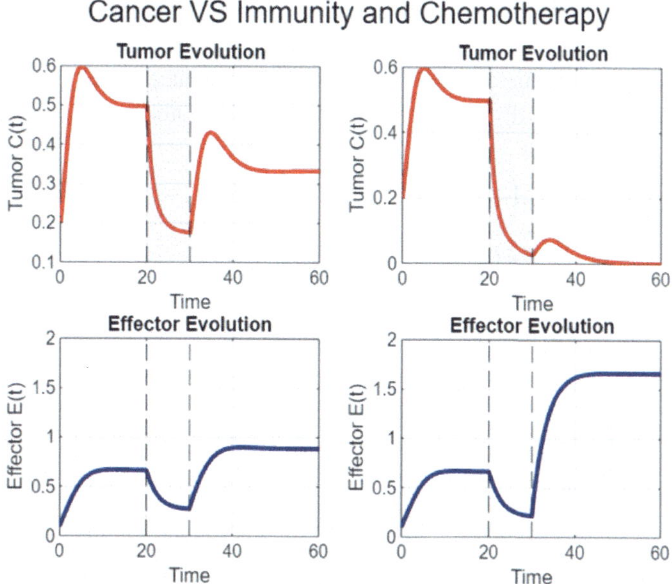

Fig. 4.6 Chemotherapy and immune response outcomes. Chemotherapy is administered during the time interval [20, 30]. The top row depicts the evolution of the tumor population $C(t)$, and the bottom row shows the effector population $E(t)$. The **left panels** show tumor recurrence, while the **right panels** show tumor elimination after chemotherapy. When the immune response is *weak*, chemotherapy reduces the tumor size during the treatment window, but the tumor eventually regrows, approaching a positive steady state. When the immune response is *strong*, the remaining tumor is cleared, and the system converges to a tumor-free steady state. Parameter values used to create this figure are given in Table 4.5

apy also suppresses the immune response while killing tumor cells. This creates a balance between tumor reduction and immune recovery. For the right panels, we observe a transient bump in tumor levels followed by decay, as the effector cell population eventually ramps up and clears the remaining tumor. When compared to surgery, which causes a sudden drop in tumor size, chemotherapy—like radiotherapy—gives the immune system time to react while treatment is happening, but it can also make the immune system weaker for a while. How these two effects balance out helps decide whether the tumor comes back or is completely removed.

Example 4.4 (Combo Therapy-Immune Model) In many cancer treatment protocols, multiple interventions are used in combination to increase the likelihood of tumor control. A typical strategy involves an initial surgery to remove the bulk of the tumor, followed by radiation and/or chemotherapy to target circulating tumor cells. Examples include treatment plans for breast cancer, colorectal cancer, and head and neck cancers, where multimodal therapy has been shown to improve long-term outcomes. We model three interventions: surgery at $t = t_s$ removes a fraction σ of the tumor mass; radiation therapy is administered over $[t_r, t_r + T_r]$ and induces additional tumor death at rate ρ; and chemotherapy is administered over $[t_c, t_c + T_c]$ and induces death of both tumor and immune cells at rates γ_C and γ_E, respectively. The resulting non-autonomous system is

$$\begin{cases} \dfrac{dC}{dt} = rC\left(1 - \dfrac{C}{K}\right) - p\,E\,C - \rho\,C\,u_r(t) - \gamma_C\,C\,u_c(t), \\ \dfrac{dE}{dt} = s + cC - d\,E - \gamma_E\,E\,u_c(t), \end{cases}$$

where $u_r(t)$ and $u_c(t)$ are indicator functions defined by

$$u_r(t) = \begin{cases} 1, & t_r \le t \le t_r + T_r, \\ 0, & \text{otherwise,} \end{cases} \qquad u_c(t) = \begin{cases} 1, & t_c \le t \le t_c + T_c, \\ 0, & \text{otherwise.} \end{cases}$$

At the time of surgery $t = t_s$, an instantaneous reduction is applied:

$$C(t_s^+) = (1 - \sigma)C(t_s^-), \qquad E(t_s^+) = E(t_s^-).$$

This example combines an impulsive event (surgery) with time-window forcing (radiation and chemotherapy). The order and overlap of treatments determine how the system can be decomposed into autonomous IVPs, which will be carried out in one of the exercises at the end of the chapter.

Remark In clinical practice, radiotherapy is typically delivered in short daily sessions, while chemotherapy exposures last much longer, so the effective treatment window for radiation is shorter than that for chemotherapy. Students should keep this time-scale difference in mind by choosing shorter forcing windows for radiotherapy than for chemotherapy when setting up or simulating treatment schedules.

4.4.4 Nonlinear Response Functions

Before we introduce **Base Tumor-Immune Model 3**, which is the Kuznetsov model, let us talk about nonlinear response functions. The goal here is to motivate the mathematical ideas behind the nonlinear terms that appear in this model and in more complex models such as the tumor-immune model in Sect. 4.5.

The Base Tumor-Immune Model 1 (exponential Lotka–Volterra) has straight nullclines and a single coexistence equilibrium. Because of this structure, the tumor-free state is never stable (it is a saddle) in that simple form. Clinically, the goal is to eliminate cancer, but in mathematical modeling we seek a framework that is as close as possible to biological reality.

The Base Tumor-Immune Model 2 adds logistic tumor growth and immune influx, making it possible to have a tumor-free equilibrium. We have seen how this simple model can be modified to include cancer treatment terms. However, it still uses a linear immune stimulation term, which means that the immune response keeps increasing without bound as the tumor grows. This is not biologically realistic.

We want to add just enough biological realism to *bend the nullclines* and *create multiple steady states*. For an undergraduate treatment of mathematical oncology these two math dynamics goals can be usually achieved through two updates:

1. replace unchecked tumor growth with logistic growth, which caps tumor size, and
2. model immune stimulation with a nonlinear saturating function..

We have implemented the first one in **Base Tumor-Immune Model 2**. Let us look into the second modification by reviewing saturating functions.

Saturating Response Functions When a stimulus x drives a response that cannot grow indefinitely, the linear term $c\,x$ eventually becomes unrealistic. A standard alternative is the Holling/Michaelis–Menten form

$$H(x) = V_{\max} \frac{x}{K + x},$$

which increases quickly at low stimulus and then levels off at $V_{\max}$. If the response turns on more sharply, a Hill function with exponent $n > 1$ can be used:

$$H_{\text{Hill}}(x; n) = V_{\max} \frac{x^n}{K^n + x^n}.$$

This introduces threshold-like behavior while staying bounded. (See [14] for a discussion on Hill functions with applications in biology.)

The Kuznetsov model [1] is a classical example of how a saturating immune stimulation term can be used in a simple but realistic tumor–immune model. After nondimensionalization, the equations take the form

$$\begin{cases} \dfrac{dC}{d\tau} = \alpha\, C\big(1 - \beta\, C\big) - E\, C, \\[2mm] \dfrac{dE}{d\tau} = \sigma + \dfrac{\rho\, E\, C}{\eta + C} - \mu\, E\, C - \delta\, E, \end{cases}$$

where C represents tumor cells and E represents effector cells. The first equation keeps the predator–prey structure but with logistic tumor growth. The second contains the nonlinear stimulation

$$\frac{\rho\, E\, C}{\eta + C} = \rho\, E \underbrace{\frac{C}{\eta + C}}_{H_{\text{Hill}}(C;\, 1)},$$

which is a Hill function with exponent $n = 1$, half-saturation $K = \eta$, and maximal rate $V_{\max} = \rho E$. For small C, $\frac{\rho EC}{\eta+C} \sim \frac{\rho}{\eta}EC$ (approximately linear); for large C, it saturates to ρE. This mathematical observation has a very important clinical application: when the tumors are small, they are handled by the immune system; but when the tumors are large, they may escape immune control when immune stimulation reaches its limit! This nonlinear saturation response function is what changes the dynamics qualitatively.

If one wishes to model a sharper response, the same term could be written as

$$\rho E \frac{C^n}{\eta^n + C^n} \qquad (n > 1),$$

which vanishes to higher order near $C = 0$ thereby delaying effective stimulation until C is close to η. This modification can serve as a starting point for an undergraduate research project.

Remark The Kuznetsov model provides a natural starting point for more detailed models that involve additional immune components such as antibodies (which is Category (2) as covered in Sect. 4.3). In our melanoma case study, Sect. 4.5, nonlinear saturating functions are used in the model.

4.4.5 Base Tumor-Immune Model 3: Kuznetsov (Classic) Model

Finally, we discuss the classic Kuznetsov model [1]. The presence of saturating response functions makes the model more complex than base models 1 and 2 and allows it to exhibit greater biological realism. It has become a standard in mathematical oncology and provides a natural starting point for advanced projects exploring immunotherapy strategies.

The paper [1] describes the development of the model, calibrated to murine lymphoma data and ultimately reduced, through non-dimensionalization and scaling, to a clean two-dimensional system:

$$\begin{cases} \dfrac{dC}{d\tau} = \alpha C(1 - \beta C) - EC, \\ \dfrac{dE}{d\tau} = \sigma + \frac{\rho EC}{\eta+C} - \mu EC - \delta E, \end{cases} \tag{4.7}$$

Consistent with our notations, $C(t)$ refers to the population of the tumor cells and $E(t)$ refers to the population of the effector cells.

The Kuznetsov model was published in 1994, when powerful visualization software were still developing. Much of the discussion in the paper was on describing the dynamics and constructing phase plane. Here, we use `ode45` to look at the phase plane of (4.7), see Fig. 4.7.

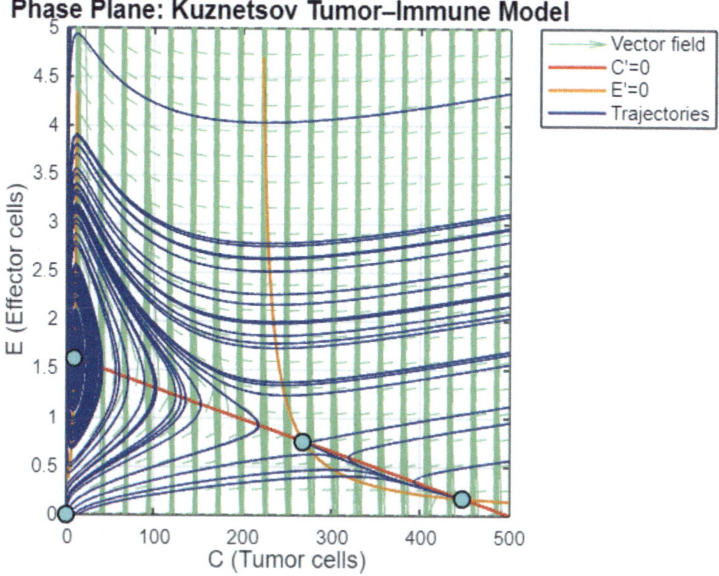

Fig. 4.7 Phase plane of the Base Tumor-Immune Model 3, (4.7). using the parameter values given in Table 4.4 in the Appendix. The system exhibits four equilibria: a tumor-free state at the origin, a spiral sink corresponding to controlled tumor dynamics, a saddle point marking unstable coexistence, and a stable node representing tumor persistence. The nullclines intersect at these equilibria. Trajectories illustrate bistability and shows that the long-term outcome depends on initial tumor and immune levels. The vector field visualizes the local direction of change in tumor and effector populations. Computations and plots were performed in MATLAB using the Kuznetsov formulation with parameters from the original study [1]

This geometry forms the foundation of many modern immunotherapy models that use ODEs. For the parameter set given in Table 4.4, there are four steady states (cyan dots in the picture):

- A *tumor-free* (trivial) equilibrium at the origin, which is a **saddle** and therefore **unstable**.
- A low-tumor, high-effector equilibrium on the left, corresponding to *tumor dormancy* (**spiral sink, stable**).
- A high-tumor, low-effector equilibrium on the right, corresponding to *tumor escape* (**nodal sink, stable**).
- A **saddle** point between the spiral and nodal sinks, whose **stable** manifold separates the basins of attraction.

We now use the Kuznetsov model to explain some clinical observations through a dynamical-systems perspective.

The tumor-free equilibrium is unstable for this parameter set; clinically, this means that immune activity alone is weak to ensure complete tumor elimination.

The dormant tumor equilibrium represents a controlled tumor under so-called immune surveillance due to being a spiral sink. The escape equilibrium corresponds to tumor dominance and immune suppression due to being a nodal sink. In dynamical systems theory, this phenomenon is called *bistability*.

The one-dimensional stable manifold of the saddle point acts as a *separatrix* dividing the phase plane into two basins of attraction. Initial conditions on one side of this threshold flow toward the dormant tumor equilibrium, while those on the other side approach the escape equilibrium where the tumor outgrows the immune system. This mathematical modeling insight, via dynamical systems, provides an explanation to a clinical observation on why small differences in initial tumor size or immune strength can lead to dramatically different long-term outcomes.

Moreover, according to Kuznetsov [1], the location of the separatrix captures the classical *sneaking through* phenomenon: a tumor slightly below the threshold remains under immune control, while a slightly larger one crosses the threshold and escapes.

Now, to close up this section, we invite students to create cancer treatment models and analyze the dynamics between cancer, immunity, and treatment strategies by modifying **Base Tumor-Immune Model 3**. This is an exciting direction because even small adjustments in treatment strength can shift the separatrix and move a patient from a tumor–escape trajectory to one of long-term immune control.

4.5 Case Study: Melanoma-Immune Competition

With some foundational insights on tumor-immune dynamics, we now turn to a real clinical context where tumor-immune interactions play a central role. Melanoma is an ideal cancer type for mathematical modeling because its progression and treatment are strongly shaped by interactions with the immune system. Melanocytes, the cells from which melanoma originates, are highly immunogenic (that is, the immune system reacts to it), and melanoma tumors often exhibit clear immune infiltration and immune escape behaviors [15]. This responsiveness makes melanoma particularly suitable for tumor–immune ODE models, and numerous mathematical frameworks have been developed specifically in this setting [16].

4.5.1 The Original Model of Pennisi

Pennisi [8] proposed a two-compartment model to describe how melanoma interacts with the immune system, see Fig. 4.8. The first compartment, the **injection site**, contains the *treatment inputs*: activated OT-1 cytotoxic T lymphocytes (CTLs) $E(t)$ and antibodies $A_b(t)$. OT-1 CTLs are laboratory-generated $CD8^+$ T cells engineered to recognize a specific target antigen, making them a convenient and well-controlled immune population for modeling. The second compartment, the **skin**, represents the *tumor microenvironment*,

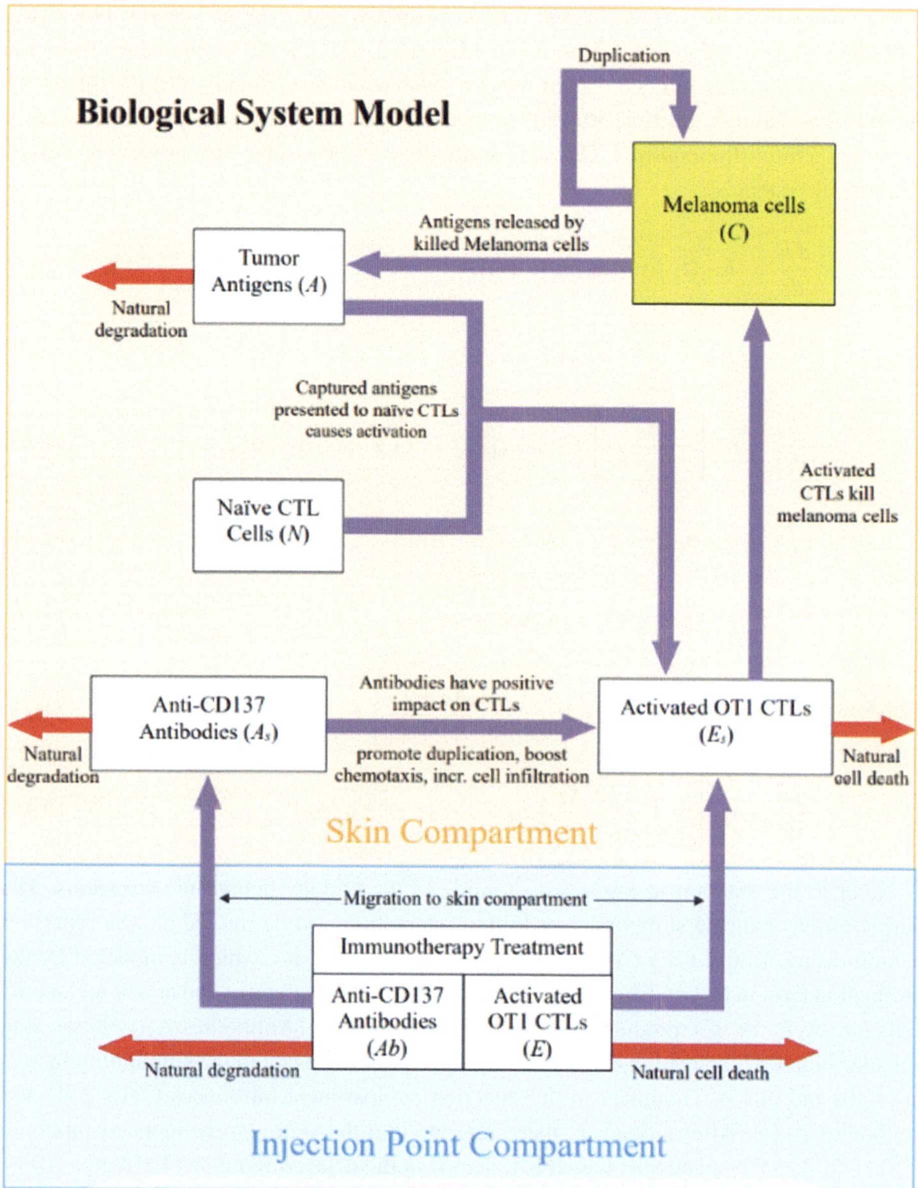

Fig. 4.8 Melanoma–immune interaction schematic. The diagram shows how tumor cells, immune effectors, and treatment agents interact. Tumor cells grow and release signals that activate immune effectors, which in turn attack the tumor. Therapeutic inputs modify these pathways by enhancing immune response or reducing tumor viability. (*Reprinted with permission from Taylor McManus and Jane Santamore. Diagram is based on the model described by Pennisi [8]*)

where melanoma cells $C(t)$ grow and interact with antigens $A(t)$, activated CTLs $E_s(t)$, antibodies $A_s(t)$, and naive CTLs $N(t)$. Migration of immune components from the injection site into the skin is modeled using a discrete delay τ. Finally, therapy inputs are incorporated through the time-varying functions $K_{\text{in}}(t, p)$ and $K_{\text{in}}(t, q)$, which specify the bolus administrations of CTLs and antibodies, respectively. The governing delay–differential system is

$$\frac{dE}{dt} = K_{in}(t, p) - \alpha_{11} E - \alpha_8 E, \tag{4.8}$$

$$\frac{dA_b}{dt} = K_{in}(t, q) - \alpha_{11} A_b - \alpha_{10} A_b, \tag{4.9}$$

$$\frac{dE_s}{dt} = \alpha_7 \left[\frac{A_s}{A_s + k_1}\right] E_s + \alpha_{11} E(t - \tau) + \alpha_6 N A - \alpha_8 E_s, \tag{4.10}$$

$$\frac{dA_s}{dt} = \alpha_{11} A_b(t - \tau) - \alpha_9 A_s E_s - \alpha_{10} A_s, \tag{4.11}$$

$$\frac{dN}{dt} = h\,(M - N) - \alpha_6 N A, \tag{4.12}$$

$$\frac{dC}{dt} = (\alpha_1 - \alpha_2 \ln C)C \; - \; \alpha_3 \left[\frac{A_s + k_2}{A_s + k_3}\right] E_s\, C, \tag{4.13}$$

$$\frac{dA}{dt} = \alpha_4 \left[\alpha_3 \left(\frac{A_s + k_2}{A_s + k_3}\right) E_s\, C\right] - \alpha_5 A - \alpha_6 N A. \tag{4.14}$$

Refer to the diagram in Fig. 4.8 as a guide to the biology behind the equations. The purple arrows indicate stimulatory or killing interactions, while the red arrows represent natural decay. Tumor cells $C(t)$ shed antigen $A(t)$ as they die, which is modeled by the production term in (4.14). Effector CTLs in the skin $E_s(t)$ recognize and attack melanoma cells (arrow $E_s \to C$), producing the killing term in (4.13). Antibodies $A_s(t)$ in the skin enhance both CTL activation and tumor killing, and this appears in the stimulation factors in (4.10) and (4.13). Treatment in the injection compartment introduces CTLs $E(t)$ and antibodies $A_b(t)$. After a delay τ, these migrate into the skin compartment as inputs to $E_s(t)$ and $A_s(t)$, respectively, which is reflected in the delayed terms in (4.10) and (4.11). The naive CTL equation also includes the term $+\alpha_6 N A$, which represents antigen-driven activation: as naive CTLs encounter antigen, they are recruited into the activation pathway that eventually contributes to the effector population in the skin.

Observe that the tumor growth term $(\alpha_1 - \alpha_2 \ln C)\,C$ used in (4.13) can be written in the standard Gompertz form shown in Table 4.1. Indeed, set $a = \alpha_2$ and $\frac{1}{b} = \exp(\frac{\alpha_1}{\alpha_2})$. So, we have

$$\alpha_1 - \alpha_2 \ln C = \alpha_2\big[\ln(1/b) - \ln C\big] = a \, \ln\left(\frac{1/b}{C}\right) = a\left(\frac{1}{bC}\right),$$

this corresponds exactly to the Gompertz growth law

$$\frac{dC}{dt} = a\,C\,\ln\left(\frac{1}{bC}\right),$$

that is, the parameterization used in the Pennisi model is mathematically equivalent to the classical Gompertz model expressed in Table 4.1.

The factors $\frac{A_s}{A_s+k_1}$ (in 4.10) and $\dfrac{A_s + k_2}{A_s + k_3}$ (in (4.13) and (4.14)), which are nonlinear saturating functions, capture antibody-driven stimulation of CTL duplication and enhancement of CTL killing, respectively. Initial conditions used in the study are

$$E(0) = A_b(0) = E_s(0) = A_s(0) = A(0) = 0, \qquad N(0) = M, \qquad C(0) = C_0.$$

This model shows how all parts of the system interact, however, it is quite complex and challenging to analyze directly. It also includes a time delay to represent cell migration, which adds another layer of mathematical difficulty. So, we next look for a simpler version that retains the essential biological interactions while replacing the delayed term by its fast-timescale approximation and focuses on the core tumor–immune feedback. Besides, the study of delay differential equations (DDEs) is much more advanced and often goes beyond the scope of an introductory modeling course.

4.5.2 Quasi-Steady-State Reduction of Fast Variables

The original Pennisi model (4.8)–(4.14) consists of seven coupled nonlinear ODEs describing the dynamics of immune effector cells, antibodies, tumor cells, antigens, and supporting immune populations. To obtain a simpler but still biologically meaningful system, we perform quasi-steady-state (QSS) reduction on the fast variables. The process yields a reduced four-equation model.

Which variables are fast? Inspection of the parameter magnitudes in Pennisi's original seven-equation melanoma model [8] confirms a clear separation of timescales between the tumor-related and immune-related variables. All rates in that model are expressed in units of Δt^{-1}, where $\Delta t = 8$ hours (see [8]). Parameters governing the tumor and antigen dynamics ($\alpha_1, \alpha_2, \alpha_4, \alpha_5$) lie between approximately 0.01 and 0.22 Δt^{-1}, corresponding to characteristic times of **several days** to **many weeks** (see Table 4.6 in the Appendix). In contrast, the local immune processes in the skin are controlled by $\alpha_6, \alpha_7, \alpha_8$, and α_{10}, whose values fall in the range 0.03–0.10 Δt^{-1}, that is, on the order of **hours** to **a few days**. In particular, $\alpha_6 = 0.10 >> h = 0.01$, $\alpha_7 = 0.095 >> \alpha_{11} = 0.009$, and $\alpha_{10} = 0.033 > \alpha_{11} = 0.009$, showing that the naive T-cell population N, the activated effectors in the skin E_s, and the antibodies in the skin A_s relax much more rapidly than the slower variables E, A_b, C, and A. This motivates treating N, E_s, and A_s as fast variables in our quasi-steady-state (QSS) reduction.

1. **QSS reduction for the naive cell population** $N(t)$. Because the naive immune population evolves on a much faster timescale than the tumor and antibody variables, it rapidly relaxes to its quasi–steady state. We therefore approximate its dynamics by setting

$$\frac{dN}{dt} \approx 0.$$

From (4.12),

$$0 = h(M - N) - \alpha_6 NA \quad \Rightarrow \quad N^*(A) = \frac{hM}{h + \alpha_6 A}.$$

Because $A(t)$ varies on a slower timescale, it is reasonable to linearize this expression around an equilibrium antigen level A^*. This yields the approximation

$$N^*(A) \approx N^*(A^*) = \frac{hM}{h + \alpha_6 A^*}.$$

which is now a constant. The term $\alpha_6 NA$, which appears in (4.10), (4.12), and (4.14), then becomes

$$\alpha_6 N^* A = \alpha_6 \left(\frac{hM}{h + \alpha_6 A^*} \right) A \approx \beta_6 A \tag{4.15}$$

where the composite parameter β_6 absorbs the constants h, M, α_6, and the reference value A^*. Thus, under this approximation, N is eliminated as a separate dynamic variable.

2. **QSS reduction for the stimulated antibody population** $A_s(t)$. The evolution equation for the skin-level antibody population A_s, Eq. (4.11), is

$$\frac{dA_s}{dt} = \alpha_{11} A_b(t - \tau) - \alpha_9 A_s E_s - \alpha_{10} A_s.$$

The migration delay τ is short (on the order of hours), while the injection-compartment antibody A_b changes on a much longer timescale. Thus, we approximate

$$A_b(t - \tau) \approx A_b(t).$$

Applying the quasi–steady-state (QSS) assumption $\frac{dA_s}{dt} \approx 0$, we solve for the steady state of A_s^* as a function of the fast variable E_s to obtain

$$0 = \alpha_{11} A_b - (\alpha_{10} + \alpha_9 E_s) A_s \quad \Rightarrow \quad A_s^* = \frac{\alpha_{11}}{\alpha_{10} + \alpha_9 E_s} A_b.$$

The factor $\dfrac{\alpha_{11}}{\alpha_{10} + \alpha_9 E_s}$ depends on the fast variable E_s. To avoid carrying this complexity through the remaining equations, we linearize around a reference value E_s^* and define the effective constant

$$k_A := \frac{\alpha_{11}}{\alpha_{10} + \alpha_9 E_s^*}.$$

Thus the stimulated antibody population is approximated by

$$A_s \approx k_A A_b, \tag{4.16}$$

and is therefore removed as an independent dynamic variable in the reduced model.

3. **QSS reduction for the stimulated effector population** $E_s(t)$. The evolution equation for the skin-level effector population E_s, Eq. (4.10), is

$$\frac{dE_s}{dt} = \alpha_7 \frac{A_s}{A_s + k_1} E_s + \alpha_{11} E(t - \tau) + \alpha_6 N A - \alpha_8 E_s.$$

The delay τ is short relative to the timescale on which $E(t)$ evolves, so we approximate $E(t - \tau) \approx E(t)$. Applying the QSS asuumption $\dfrac{dE_s}{dt} \approx 0$ and solving for E_s gives

$$E_s^* = \frac{\alpha_{11} E + \alpha_6 N A}{\alpha_8 - \alpha_7 \dfrac{A_s}{A_s + k_1}}.$$

Next, we substitute the QSS expressions already derived for the other fast variables. Using (4.15)

$$N \approx N^* \qquad \text{and} \qquad \alpha_6 N^* A \approx \beta_6 A,$$

together with (4.16)

$$A_s \approx k_A A_b,$$

we obtain

$$E_s^* = \frac{\alpha_{11} E + \beta_6 A}{\alpha_8 - \alpha_7 \dfrac{k_A A_b}{k_A A_b + k_1}}.$$

The denominator still depends on the fast variable A_s (through $k_A A_b$). Because both A_s and E_s adjust rapidly to their quasi–steady values, this denominator is effectively constant on the slower E, A_b, C, and A timescales. To avoid introducing additional nonlinearities and parameter dependence into the reduced model, we approximate

$$E_s \approx k_E\big(\alpha_{11}E + \beta_6 A\big), \tag{4.17}$$

where

$$k_E := \left(\alpha_8 - \alpha_7 \frac{k_A A_b^*}{k_A A_b^* + k_1}\right)^{-1}$$

is an effective parameter. Thus, under this QSS approximation, E_s is no longer treated as an independent dynamic variable in the reduced model.

4. **Focusing on the antigen.** To obtain a closed reduced system, we must eliminate A_s and E_s from the interaction terms in (4.13) and (4.14), where they enter through the Hill-type expression

$$\alpha_3 \frac{A_s + k_2}{A_s + k_3} E_s C.$$

Substituting the QSS approximations (4.16) and (4.17), this becomes

$$\alpha_3 \frac{k_A A_b + k_2}{k_A A_b + k_3} k_E\big(\alpha_{11}E + \beta_6 A\big)C.$$

To keep the reduced model tractable, we replace this complicated dependence on the fast variables A_s and E_s by a simpler Hill-type function of the slower antigen variable A:

$$G(A) = \frac{A + k_2}{A + k_3}.$$

Thus we approximate

$$\alpha_3 \frac{k_A A_b + k_2}{k_A A_b + k_3} k_E\big(\alpha_{11}E + \beta_6 A\big)C \;\approx\; \beta_3 \frac{A + k_2}{A + k_3} EC, \tag{4.18}$$

where the effective parameter β_3 depends on α_3, k_A, and k_E, and preserves the original Hill-type interaction structure.

Remark There is also another way to do this approximation. Instead of using the antigen A, we could base the Hill function on the antibody A_b and get a different reduced model. See Problem 4.13 at the end of the chapter.

5. **Rewriting the tumor and antigen equations.** The tumor equation (4.13) originally used a Gompertz growth term and included the fast variables E_s and A_s in the immune-mediated killing term. For analytical tractability, we replace the Gompertz law with

a logistic growth term and substitute the QSS approximations for the fast variables, namely (4.16) and (4.17), together with the interaction approximation (4.18). This yields the reduced tumor dynamics

$$\frac{dC}{dt} = \alpha_1 C \left(1 - \frac{C}{K}\right) - \beta_3 \frac{A + k_2}{A + k_3} EC.$$

Similarly, substituting the QSS expressions into the antigen equation (4.14) gives

$$\frac{dA}{dt} = \alpha_4 \beta_3 \frac{A + k_2}{A + k_3} EC - (\alpha_5 + \beta_6) A.$$

6. **Simplifying the immune equations.** The effector equation (4.8) is

$$\frac{dE}{dt} = K_{in}(t, p) - \alpha_{11} E - \alpha_8 E,$$

while the antibody equation (4.9) is

$$\frac{dA_b}{dt} = K_{in}(t, q) - \alpha_{11} A_b - \alpha_{10} A_b.$$

First, we observe that there are two clearance terms in both equations and hence, to simplify, we introduce new effective parameters:

$$\beta_8 = \alpha_{11} + \alpha_8, \qquad \beta_{10} = \alpha_{11} + \alpha_{10}.$$

Hence, the effector equation is now

$$\frac{dE}{dt} = p - \beta_8 E,$$

and the antibody equation is now

$$\frac{dA_b}{dt} = q - \beta_{10} A_b.$$

Here, we assume that immune stimulation is constant over the timescale of interest and therefore replace functions $K_{in}(t, p)$ and $K_{in}(t, q)$ by constants p and q.

Combining these reductions and approximations yields the simplified system:

$$\frac{dE}{dt} = p - \beta_8 E, \tag{4.19}$$

$$\frac{dA_b}{dt} = q - \beta_{10} A_b, \tag{4.20}$$

$$\frac{dC}{dt} = \alpha_1 C\left(1 - \frac{C}{K}\right) - \beta_3 \frac{A + k_2}{A + k_3} EC, \tag{4.21}$$

$$\frac{dA}{dt} = \alpha_4 \beta_3 \frac{A + k_2}{A + k_3} EC - (\alpha_5 + \beta_6)A. \tag{4.22}$$

The four-equation system (4.19)–(4.22) represents a tractable four–dimensional reduction of the full melanoma–immune model obtained by applying quasi–steady-state (QSS) approximations to the fast skin-level variables and simplifying the original interaction terms. Because several modeling choices were made during this reduction let us state these assumptions explicitly.

4.5.3 Modeling Assumptions

The reduced system of four ODEs (4.19)–(4.22) is derived under the following assumptions:

1. **Constant stimulation of immune components.** Effector cells receive a constant baseline input p and decay at an effective rate β_8. Similarly, antibodies receive a constant baseline input q and decay at an effective rate β_{10}.
2. **Logistic tumor growth.** Tumor cells grow according to a logistic law

$$\alpha_1 C\left(1 - \frac{C}{K}\right),$$

which replaces the original Gompertz growth term for analytical tractability.
3. **Antigen-modulated immune killing.** Effector-mediated killing of tumor cells occurs through the Hill-type interaction term

$$-\beta_3 \frac{A + k_2}{A + k_3} EC.$$

4. **Antigen production and clearance.** Antigen is produced when effector cells interact with tumor cells, represented by

$$\alpha_4 \beta_3 \frac{A + k_2}{A + k_3} EC,$$

and is removed via natural decay $-\alpha_5 A$ and an additional immune-mediated clearance term $-\beta_6 A$.
5. **Fast–slow timescale separation.** The skin-level populations N, A_s, and E_s evolve on much faster timescales than the tumor variables and injection-compartment immune

variables. Thus their time derivatives are approximated by zero and their QSS values are used in the reduced model.

6. **Delay approximation.** The migration delay τ between compartments is short relative to the dynamics of E and A_b, allowing the approximation

$$E(t-\tau) \approx E(t), \qquad A_b(t-\tau) \approx A_b(t).$$

4.5.4 Equilibrium and Local Stability

After reducing the full seven-equation Pennisi to a four-equation model, we now study the long-term behavior of the model. Equilibrium points play a central role in understanding the possible biological outcomes of the tumor–immune interaction: complete tumor elimination, persistent controlled tumor, or uncontrolled growth.

Equilibrium Equations An equilibrium point

$$X^* = (E^*, A_b^*, C^*, A^*)$$

satisfies

$$\frac{dE}{dt} = \frac{dA_b}{dt} = \frac{dC}{dt} = \frac{dA}{dt} = 0.$$

Using the reduced system (4.19)–(4.22), the equilibrium conditions become

$$E^* = \frac{p}{\beta_8}, \qquad A_b^* = \frac{q}{\beta_{10}}, \tag{4.23}$$

together with the coupled nonlinear equations

$$\alpha_1 C^* \left(1 - \frac{C^*}{K}\right) = \beta_3\, G(A^*)\, E^* C^*, \qquad \alpha_4 \beta_3\, G(A^*)\, E^*\, C^* = (\alpha_5 + \beta_6)\, A^*,$$

where $G(A) = \dfrac{A + k_2}{A + k_3}$. These determine the tumor and antigen components of the equilibrium.

Tumor-Free Equilibrium Setting $C^* = 0$ in the reduced system forces $A^* = 0$ in the antigen equation. The immune equations then give

$$E^* = \frac{p}{\beta_8}, \qquad A_b^* = \frac{q}{\beta_{10}}. \tag{4.24}$$

Thus the tumor-free equilibrium is

$$X_{\mathrm{TF}}^* = \left(\frac{p}{\beta_8},\ \frac{q}{\beta_{10}},\ 0,\ 0\right).$$

Local Stability To study its stability, we linearize the system about X_{TF}^*. Recall that $G(A) = \dfrac{A + k_2}{A + k_3}$, so at the tumor-free state ($A^* = 0$) we have

$$G(0) = \frac{k_2}{k_3}.$$

The Jacobian at X_{TF}^* is block lower–triangular:

$$J(X_{\mathrm{TF}}^*) = \begin{pmatrix} -\beta_8 & 0 & 0 & 0 \\ 0 & -\beta_{10} & 0 & 0 \\ 0 & 0 & \alpha_1 - \beta_3 G(0) E^* & 0 \\ 0 & 0 & \alpha_4 \beta_3 G(0) E^* & -(\alpha_5 + \beta_6) \end{pmatrix}.$$

Three of the four eigenvalues are clearly negative:

$$\lambda_1 = -\beta_8, \qquad \lambda_2 = -\beta_{10}, \qquad \lambda_4 = -(\alpha_5 + \beta_6).$$

The remaining eigenvalue,

$$\lambda_3 = \alpha_1 - \beta_3 G(0) E^*,$$

determines stability of the tumor-free state.

Theorem 4.1 (Local Stability of the Tumor-Free Equilibrium) *The tumor-free equilibrium*

$$X_{\mathrm{TF}}^* = \left(\frac{p}{\beta_8},\ \frac{q}{\beta_{10}},\ 0,\ 0\right)$$

is locally asymptotically stable if and only if

$$\beta_3 G(0)\, E^* > \alpha_1, \tag{4.25}$$

where $G(0) = \dfrac{k_2}{k_3}$. *That is, the effective immune killing rate* $\beta_3 G(0) E^*$ *must exceed the intrinsic tumor growth rate* α_1. □

Biologically, this means that when the immune response is strong enough at baseline, small tumor populations are driven to extinction rather than allowed to grow.

Tumor-Persistent Equilibrium There is a second equilibrium. Recall that E^* and A_b^* are given by (4.24). A positive-tumor equilibrium has $C^* > 0$ and $A^* > 0$ and must satisfy

$$\alpha_1\left(1 - \frac{C^*}{K}\right) = \beta_3\, G(A^*)\, E^*, \qquad \alpha_4\beta_3\, G(A^*)\, E^*\, C^* = (\alpha_5 + \beta_6)\, A^*.$$

Solving the first equation for C^* gives

$$C^* = K\left[1 - \frac{\beta_3}{\alpha_1} G(A^*)E^*\right].$$

A biologically feasible equilibrium therefore requires

$$\beta_3\, G(A^*)\, E^* < \alpha_1.$$

Theorem 4.2 (Feasibility of a Tumor-Persistent Equilibrium) *A coexistence equilibrium with $C^* > 0$ and $A^* > 0$ can exist only if*

$$\beta_3\, G(A^*)\, E^* < \alpha_1.$$

This expresses a simple idea: a persistent tumor state is possible only when immune killing is not strong enough to suppress intrinsic tumor growth α_1.

Comparison of the Two Regimes The two inequalities describing the tumor-free and tumor-persistent regimes are:

$$\begin{cases} \beta_3 G(0)E^* > \alpha_1 & \Rightarrow \text{Tumor-free state is stable,} \\ \beta_3 G(A^*)E^* < \alpha_1 & \Rightarrow \text{Tumor persists at a positive level.} \end{cases}$$

These are not algebraic opposites, because they apply at different equilibria and because the antigen-weighting factor $G(A^*)$ appears only in the tumor-persistent case. They summarize a single biological theme: either the immune system dominates (tumor cleared) or it is too weak (tumor persists).

Monostability Taken together, the tumor-free and tumor-persistent inequalities define a single threshold that separates two distinct long-term outcomes. This leads to the following result.

Theorem 4.3 (Monostability of the Reduced Melanoma–Immune System) *Assume all parameters are nonnegative and $k_2 > k_3 > 0$. Under the conditions stated in Theorems 4.1 and 4.2, the reduced system admits at most one asymptotically stable equilibrium in the nonnegative orthant.*

Proof For $G(A) = \dfrac{A + k_2}{A + k_3}$ with $k_2 > k_3 > 0$, the function G is strictly increasing on $A \geq 0$. Thus, whenever $A^* > 0$,

$$G(A^*) > G(0) = \frac{k_2}{k_3}.$$

The tumor-free equilibrium is stable only if

$$\beta_3 \, G(0) \, E^* > \alpha_1,$$

while a tumor-persistent equilibrium can exist only if

$$\beta_3 \, G(A^*) \, E^* < \alpha_1.$$

Because $G(A^*) > G(0)$, these two inequalities cannot hold simultaneously for the same parameter set. Therefore, the tumor-free and tumor-persistent equilibria cannot both be stable. All trajectories approach either the tumor-free equilibrium or a single tumor-persistent equilibrium. □

This property is helpful for analysis and for student projects: the reduced model exhibits a single threshold determined by the competition between immune strength and tumor growth. Students who want to go further can explore how small changes to the stimulation or killing terms in extended models may lead to bistability.

4.5.5 Non-negativity and Invariance of the Biologically Feasible Region

Before analyzing the stability of the tumor-free equilibrium, we must first confirm that the model respects a basic biological requirement: cell populations cannot become negative. The natural way to formalize this is to show that the nonnegative orthant is forward invariant under the dynamics. This guarantees that all trajectories starting with biologically meaningful initial data remain biologically meaningful for all future time.

Theorem 4.4 (Forward Invariance of the Biologically Relevant Region) *Let*

$$\Omega = \{(E, A_b, C, A) \in \mathbb{R}^4 : E \geq 0, \; A_b \geq 0, \; C \geq 0, \; A \geq 0\}.$$

Consider the reduced tumor-immune model (4.19)–(4.22)*, with all parameters nonnegative. Then* Ω *is forward invariant: if the initial condition lies in* Ω*, the solution remains in* Ω *for all future time, that is,*

$$(E(0), A_b(0), C(0), A(0)) \in \Omega \quad \Longrightarrow \quad (E(t), A_b(t), C(t), A(t)) \in \Omega \textit{ for all } t \geq 0.$$

We verify that the vector field does not point outward on any coordinate hyperplane. The standard procedure (see [17, 18]) is:

1. Set one variable equal to zero while keeping the others nonnegative.
2. Evaluate its time derivative on that boundary.
3. Show that the derivative is nonnegative.

Proof *Effector Variable.* At $E = 0$ (with $A_b, C, A \geq 0$),

$$\left.\frac{dE}{dt}\right|_{E=0} = p \geq 0.$$

Thus E cannot cross into negative values.
Antibody Variable. At $A_b = 0$ (with $E, C, A \geq 0$),

$$\left.\frac{dA_b}{dt}\right|_{A_b=0} = q \geq 0.$$

Thus $A_b \geq 0$ is preserved.
Tumor Variable. At $C = 0$,

$$\left.\frac{dC}{dt}\right|_{C=0} = \alpha_1 \cdot 0 - \beta_3 G(A)E \cdot 0 = 0.$$

Hence the plane $C = 0$ is invariant.
Antigen Variable. At $A = 0$, using $G(0) = k_2/k_3$,

$$\left.\frac{dA}{dt}\right|_{A=0} = \alpha_4 \beta_3 G(0) EC \geq 0.$$

Since the vector field is inward-pointing or tangent on every boundary of Ω, the region is forward invariant. □

Forward invariance ensures that all trajectories starting in Ω remain in the biologically feasible region. This will allow the stability estimates in the next subsection to hold for all $t \geq 0$.

4.5.6 Global Stability of the Tumor-Free Equilibrium

Now that solutions are known to remain in Ω, we analyze the long-term behavior of the system. Our goal is to show that the tumor-free equilibrium X^*_{TF} is not only locally stable but in fact attracts every trajectory in Ω. The proof uses two complementary ideas: a Lyapunov function controls the immune variables E and A_b, while the tumor and antigen variables require separate comparison estimates coming directly from their differential equations.

Definition 4.1 (Global Asymptotic Stability) Let x^* be an equilibrium of $\dot{x} = f(x)$. We say that x^* is *globally asymptotically stable* in a region Ω if:

1. **Stability.** For every $\varepsilon > 0$ there exists $\delta > 0$ such that

$$\|x(0) - x^*\| < \delta \quad \Longrightarrow \quad \|x(t) - x^*\| < \varepsilon \text{ for all } t \geq 0.$$

2. **Global attractivity.** Every solution starting in Ω converges to x^*:

$$x(0) \in \Omega \quad \Longrightarrow \quad \lim_{t\to\infty} x(t) = x^*.$$

Although forward invariance is not mentioned in the definition, it is essential in practice: to prove global attractivity on Ω, we must guarantee that trajectories do not leave Ω before convergence can occur.

The reduced model (4.19)–(4.22) has the tumor-free equilibrium

$$X^*_{TF} = (E^*, A^*_b, 0, 0), \qquad E^* = \frac{p}{\beta_8}, \qquad A^*_b = \frac{q}{\beta_{10}},$$

and it is locally asymptotically stable whenever

$$\beta_3 G(0) E^* > \alpha_1, \qquad G(0) = \frac{k_2}{k_3}.$$

We now show that the same condition guarantees global asymptotic stability but first we prove a lemma which will be need in the proof.

Lemma 4.1 (Explicit Solutions for the Immune Variables) *The effector and antibody equations (4.19) and (4.20)*

$$\frac{dE}{dt} = p - \beta_8 E, \qquad \frac{dA_b}{dt} = q - \beta_{10} A_b$$

have the explicit solutions

$$E(t) = E^* + \big(E(0) - E^*\big)e^{-\beta_8 t}, \qquad A_b(t) = A_b^* + \big(A_b(0) - A_b^*\big)e^{-\beta_{10} t}, \tag{4.26}$$

where $E^* = \dfrac{p}{\beta_8}$ *and* $A_b^* = \dfrac{q}{\beta_{10}}$. *Hence,* $E(t) \to E^*$ *and* $A_b(t) \to A_b^*$ *exponentially fast as* $t \to \infty$.

Proof Rewrite the effector equation as

$$\frac{dE}{dt} + \beta_8 E = p.$$

Subtracting the equilibrium $E^* = p/\beta_8$ gives

$$\frac{d}{dt}(E - E^*) = -\beta_8(E - E^*).$$

Define $u(t) = E(t) - E^*$. Then u satisfies the linear differential equation

$$\dot{u} = -\beta_8 u,$$

which has the solution

$$u(t) = u(0)e^{-\beta_8 t}.$$

Thus

$$E(t) = E^* + \big(E(0) - E^*\big)e^{-\beta_8 t}.$$

The derivation for $A_b(t)$ is analogues. □

Theorem 4.5 (Global Asymptotic Stability of the Tumor-Free Equilibrium) *Assume all parameters are nonnegative and* $(E(0), A_b(0), C(0), A(0)) \in \Omega$. *If* $\beta_3 G(0) E^* > \alpha_1$, *then the tumor-free equilibrium*

$$X_{TF}^* = (E^*, A_b^*, 0, 0)$$

is globally asymptotically stable in Ω.

Proof Consider the Lyapunov function

$$V(E, A_b, C, A) = \frac{1}{2}\big[(E - E^*)^2 + (A_b - A_b^*)^2 + C^2 + A^2\big],$$

which is positively definite with respect to the equilibrium $X^* = (E^*, A_b^*, 0, 0)$. The derivative of V along solutions of the reduced model is

$$\dot{V} = (E - E^*)\frac{dE}{dt} + (A_b - A_b^*)\frac{dA_b}{dt} + C\frac{dC}{dt} + A\frac{dA}{dt}.$$

Substituting the derivatives from (4.19)–(4.22) and using the equilibrium relations $p = \beta_8 E^*$ and $q = \beta_{10} A_b^*$ gives

$$\dot{V} = -\beta_8(E - E^*)^2 - \beta_{10}(A_b - A_b^*)^2 +$$
$$+ C^2\left(\alpha_1 - \alpha_1\frac{C}{K} - \beta_3 G(A)E\right) + A\left(\alpha_4\beta_3 G(A)EC - \left(\alpha_5 + \beta_6\right)A\right).$$

Immune Variables. The first line is strictly nonpositive and vanishes only at $E = E^*$ and $A_b = A_b^*$. Thus the immune variables satisfy the stability condition in the definition of GAS.

Tumor Variable. The Lyapunov derivative is not sign-definite in C, so we analyze the tumor equation directly. From (4.21),

$$\frac{1}{C}\frac{dC}{dt} = \alpha_1\left(1 - \frac{C}{K}\right) - \beta_3 G(A)E \le \alpha_1 - \beta_3 G(A)E.$$

Because $E(t)$ converges exponentially to E^* as shown in Lemma 4.1 and $A(t) \ge 0$ for all t, the function $G(A) = \frac{A+k_2}{A+k_3}$ satisfies $G(A(t)) \ge G(0)$ (assuming $k_3 > k_2 > 0$). Therefore the inequality $\beta_3 G(0)E^* > \alpha_1$ implies that there exist T_1 and $\gamma_C > 0$ such that

$$\alpha_1 - \beta_3 G(A(t))E(t) \le -\gamma_C \qquad \text{for all } t \ge T_1.$$

Hence, for $t \ge T_1$,

$$\frac{1}{C}\frac{dC}{dt} \le -\gamma_C.$$

Integrating this inequality from T_1 to t gives

$$\ln C(t) - \ln C(T_1) \le -\gamma_C(t - T_1),$$

so

$$C(t) \le C(T_1)e^{-\gamma_C(t-T_1)} \to 0 \qquad \text{as } t \to \infty.$$

Thus $C(t)$ decays to zero.

Antigen Variable. Once $C(t)$ is small, the forcing term in the antigen equation becomes small. From (4.22),

$$\frac{dA}{dt} = \alpha_4\beta_3 G(A)EC - (\alpha_5 + \beta_6)A.$$

Rewrite this as

$$\frac{dA}{dt} + (\alpha_5 + \beta_6)A = r(t), \qquad r(t) = \alpha_4\beta_3 G(A(t))E(t)C(t).$$

Because $G(A)$ and $E(t)$ are bounded and $C(t) \to 0$, we have $r(t) \to 0$. Variation of parameters gives

$$A(t) = A(T_2)e^{-(\alpha_5+\beta_6)(t-T_2)} + \int_{T_2}^{t} e^{-(\alpha_5+\beta_6)(t-s)} r(s)\, ds,$$

which tends to zero because the kernel is integrable and $r(s) \to 0$.

Together, the Lyapunov argument ensures stability at X^*_{TF}, and the decay of $C(t)$ and $A(t)$ shows that every trajectory in Ω converges to the equilibrium. Thus both conditions in the definition of global asymptotic stability hold. □

Biological Interpretation of the Stability Result This result is for the reduced model obtained after the QSS step. The reduced system keeps the key interactions but does not include all of the immune system's complexity. The condition $\beta_3 G(0)E^* > \alpha_1$ means that the immune system's basic killing strength is stronger than the tumor's natural growth rate. When this holds, the structure of the equations forces the tumor to decline for every biologically meaningful initial condition. It may feel surprising biologically that the initial tumor size does not matter, but this is a feature of the simplified dynamics. Remember that this came from our reduced model obtained via QSS approximations! In this regime, the tumor decreases, antigen production stops, and the immune variables return to their steady values. This also shows where the model can be extended if students want to add more realistic behavior, see Hallmarks of Cancer in Sect. 4.7. Finally, what we have obtained here should not be used for clinical prediction.

4.6 Numerical Simulations

This section presents numerical simulations of the reduced model (4.19)–(4.22) to illustrate analytical results established in Theorems 4.1 and 4.2. The simulations demonstrate two biologically distinct steady-state outcomes: a tumor-free equilibrium, in which the immune response eliminates the tumor, and a tumor-persistent equilibrium, in which the tumor survives at a small but positive level. The only difference between the two cases lies in the immune killing parameter β_3, found in (4.21) and (4.22), which controls the strength of the effector-mediated tumor response. For each case, we produce visualizations of the evolution plots (or time series) and the phase plane plots. Table 4.7 summarizes all variables, parameters, and values used in the simulations.

The effector and antibody equations are 1st-order linear ODEs, $E' = p - \beta_8 E$ and $A_b' = q - \beta_{10} A_b$, with steady states $E^* = p/\beta_8 = 1/3$ and $A_b^* = q/\beta_{10} = 0.2$. In our simulations $A_b(0) = A_b^*$, so the antibody solution is simply $A_b(t) \equiv A_b^*$ and remains constant throughout each run.

Tumor-Free Equilibrium Figure 4.9 shows the time evolution of the four state variables on $t \in [0, 100]$. Because the effector and antibody equations are 1st-order linear ODEs, $E(t)$ and $A_b(t)$ converge to their equilibrium values $E^* = p/\beta_8$ and $A_b^* = q/\beta_{10}$. The tumor and antigen dynamics agree with the stability analysis. The killing function satisfies $G(0) = k_2/k_3 = 4/3$, so the immune killing rate at equilibrium is

$$\beta_3 G(0) E^* = 2.5 \cdot (4/3) \cdot (1/3) = 10/9 \approx 1.1111,$$

which exceeds the intrinsic growth rate $\alpha_1 = 1$. Thus the stability condition $\beta_3 G(0) E^* > \alpha_1$ holds. Consistent with this, the tumor level $C(t)$ decreases monotonically to zero. Antigen concentration $A(t)$ shows a small initial rise from early tumor killing and then decays rapidly once antigen production slows. Both $C(t)$ and $A(t)$ converge smoothly to zero, confirming global asymptotic stability of the tumor-free equilibrium.

Figure 4.9 also shows the phase–plane plots for (C, A) and (E, C). In the (C, A) plane, the curve starts at (0.6, 0.2) (black dot), rises slightly as antigen is released during tumor killing, and then both variables decrease to zero; cancer and antigen are cleared. In the (E, C) plane, the trajectory begins at (0.5, 0.6) (black dot). By the tumor-free stability theorem, $C(t)$ decreases to zero, so the curve moves downward as the tumor dies out and leftward as $E(t)$ decreases toward E^*. The solution ends at the tumor-free equilibrium (red dot).

Tumor-Persistent Equilibrium Figure 4.10 shows the time evolution of the four state variables on $t \in [0, 200]$ (a longer window than the tumor-free case). Since $E(t)$ and $A_b(t)$ evolve according to the linear equations described at the beginning of this section, their behavior is already determined by the steady states $E^* = p/\beta_8 = 1/3$ and $A_b^* = q/\beta_{10} = 0.2$.

To assess whether a positive tumor equilibrium can occur, we apply Theorem 4.2, which states that a coexistence equilibrium with $C^* > 0$ and $A^* > 0$ can exist only if

$$\beta_3 \, G(A^*) \, E^* < \alpha_1.$$

In this simulation the antigen steady state is $A^* \approx 0.1617$. Using this value in the killing function

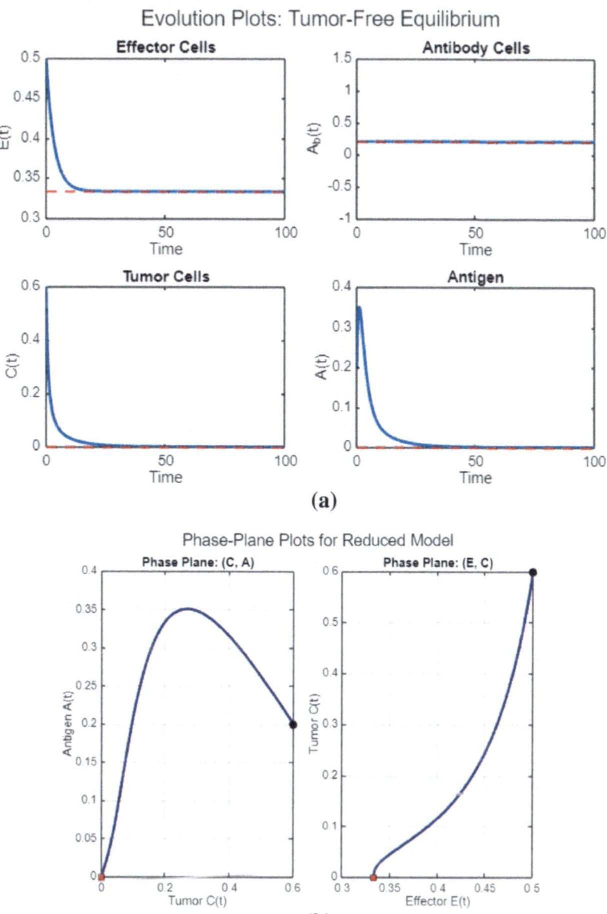

Fig. 4.9 Numerical simulations for the reduced tumor-immune model, tumor-free case. The time-series panels on the top show convergence of effector E and antibody A_b to their steady-states, while tumor C and antigen A levels decay to zero. Phase plane trajectories show a brief increase in antigen in (C, A) and a monotone decay in (E, C); read from the black dot (initial condition) to the red dot (tumor-free equilibrium). (**a**) Time-series plots for (E, A_b, C, A). (**b**) Phase plane plots on (C, A) and (E, C). Parameter values for these numerical simulations are given in Table 4.7 in the Appendix

$$G(A^*) = \frac{A^* + k_2}{A^* + k_3} = \frac{0.1617 + 0.4}{0.1617 + 0.3} \approx 1.2165,$$

the effective immune–killing rate in Theorem 4.2 becomes

$$\beta_3 G(A^*) E^* = 0.5 \cdot (1.2165) \cdot \frac{1}{3} \approx 0.2028 < \alpha_1 = 1.$$

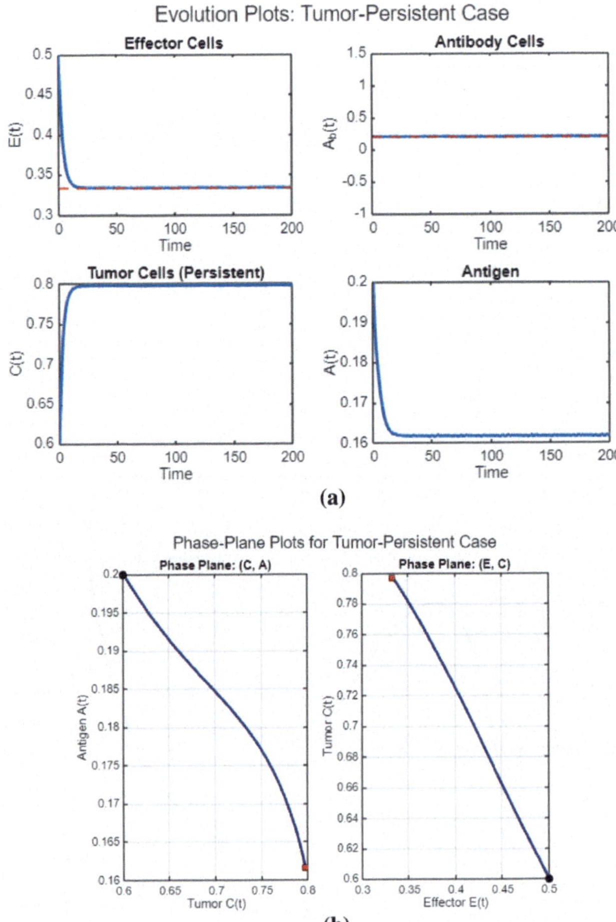

Fig. 4.10 Numerical simulations for the reduced tumor-immune model, tumor-persistent case. The time-series plots show convergence of effector E and antibody A_b to their steady-states, while tumor C approaches a positive persistent level and antigen A decays to zero. Phase-plane trajectories show a monotone rise in tumor burden in both (C, A) and (E, C); the black dot marks the initial condition and the red dot marks the tumor-persistent equilibrium. (**a**) Time-series plots for (E, A_b, C, A). (**b**) Phase plane plots on (C, A) and (E, C). Parameter values for these numerical simulations are given in Table 4.7 in the Appendix

Thus the condition in Theorem 4.2 is satisfied, and a tumor-persistent equilibrium is feasible. The numerical solution reflects this: $C(t)$ moves away from 0 and settles at a positive steady value C^*, while $A(t)$ approaches the small positive level $A^* \approx 0.162$.

The phase–plane plots show the same dynamic story. In the (C, A) plane, the trajectory begins at $(0.6, 0.2)$ (black dot) and moves toward the equilibrium $(C^*, A^*) > (0, 0)$. In the (E, C) plane, the curve starts near $(0.5, 0.6)$ and drifts leftward and downward as $E(t)$ approaches E^* and $C(t)$ converges to C^*, ending at the tumor-persistent equilibrium (red dot).

4.7 Hallmarks of Cancer: Linking Biology and Mathematical Model

At some point in the mathematical modeling experience, the student will need to communicate their results to a biologist or oncologist. This step is both challenging and rewarding—it requires translating mathematical symbols and equations into biological meaning. The framework of *Hallmarks of Cancer*, introduced by Hanahan and Weinberg in 2000 [6], provides a language that can bridge this gap between disciplines. By connecting each mathematical term to a biological capability, the students gain a deeper understanding of what the equations represent to better communicate tumor-immune dynamics.

The original six hallmarks summarize the strategies that tumor cells use to grow, survive, and spread. We briefly state the six hallmarks here:

- **Sustaining proliferative signaling:** Cancer cells keep receiving (or even producing) signals that tell them to divide and grow, when normal cells would stop.
- **Evading growth suppressors:** Cancer cells ignore regulatory signals that would slow or stop growth.
- **Activating invasion and metastasis:** Cancer cells acquire the ability to move and establish new colonies of cancer cells.
- **Enabling replicative immortality:** Cancer cells maintain their ability to divide indefinitely.
- **Inducing angiogenesis:** Cancer cells promote new blood vessel growth to supply nutrients (and hence continue expansion).
- **Resisting cell death:** Cancer cells avoid programmed cell death, also known as apoptosis, which removes cells that are old, damaged, or no longer needed.

Table 4.2 shows a mapping between the terms in the 2D tumor-immune model with treatments and the six hallmarks.

In 2011, Hanahan and Weinberg [19] updated the original six hallmark capabilities to include two emerging hallmarks and two enabling characteristics of cancer growth:

- **Deregulating cellular energetics:** Cancer cells alter how they make and use energy that favors rapid growth.
- **Avoding immune destruction:** Cancer cells evade and avoid detection of the immune system.
- **Genome instability and mutation:** Cancer cells develop genetic adaptation to stress or therapy.
- **Tumor-promoting inflammation:** Chronic inflammation supports tumor growth and survival.

Problem 4.14 at the end of the section invites the reader to map the full Pennisi model (4.8)–(4.14) and the reduced model (4.19)–(4.22) to the updated 2011 hallmarks.

Table 4.2 Mapping the Base Tumor-Immune Model 2 to the six original hallmarks of cancer [6]. Each term in the two-equation system represents a biological process linked to a hallmark capability. There are some non-hallmark terms

Mapping the base tumor-immune model 2 with the six hallmarks	
$\frac{dC}{dt} = rC\left(1 - \frac{C}{K}\right) - pEC - T_C(t)$, $\frac{dE}{dt} = s + cC - dE$	
$rC(1 - C/K)$	**Sustaining proliferative signaling** **Evading growth suppressors**
$-pEC$	**Resisting cell death** Direct encounter cancer killing
$-T_C(t)$	**Treatment effects** (surgery, radiation, chemotherapy) *Not a hallmark*
s	**Enabling replicative immortality** Constant immune source
cC	**Tumor-driven immune activation** Immune stimulation by tumor presence *Not a hallmark*
$-dE$	**Effector decay/exhaustion** Loss of immune surveillance *Not a hallmark*

Now, the reduced model (4.19)–(4.22) captures the main feedback loop between tumor cells C, effector cells E, antibodies A_b, and antigen A. The QSS method was used in this process. Even with the simplified model, the terms already reflect several major biological ideas, such as immune killing of tumor cells, logistic tumor growth, and antigen production from tumor turnover.

Here, we enrich the model biologically by adding some terms. These additions lead to the extended reduced model:

$$
\begin{aligned}
\frac{dE}{dt} &= p + \underbrace{\alpha_7 \frac{A}{A+k_1} E}_{\text{antigen-driven effector stimulation}} + \underbrace{\beta_6 A}_{\text{antigen-mediated boost}} - \alpha_8 E, \\
\frac{dA_b}{dt} &= q + \underbrace{\alpha_{11} E}_{\text{effector-driven antibody production}} - \alpha_{10} A_b, \\
\frac{dC}{dt} &= \alpha_1 C\left(1 - \frac{C}{K}\right) - \alpha_3 \frac{A+k_2}{A+k_3} EC, \\
\frac{dA}{dt} &= \alpha_4 \alpha_3 \frac{A+k_2}{A+k_3} EC - (\alpha_5 + \beta_6) A - \underbrace{\alpha_{12} A_b A}_{\text{antibody-assisted antigen clearance}} .
\end{aligned}
\tag{4.27}
$$

The new terms represent specific processes rather than general complexity. For example, the expression $H(A) = \alpha_7 \frac{A}{A+k_1} E$ models antigen-driven recruitment or activation of effector cells. This captures the idea that rising antigen levels help trigger and amplify the immune response, but only up to a saturation level when the system can no longer increase activation. Other nonlinear saturation responses can be used to capture this item. The additional term $\beta_6 A$ represents effector production that occurs when antigen levels increase. The term $\alpha_{11} E$ in the antibody equation represents antibody production, say, during an inflammatory response. The term $-\alpha_{12} A_b A$ in the antigen equation describes antibody-assisted antigen removal.

The terms in the extended reduced model (4.27) are mapped to the 2011 hallmarks in Table 4.3.

4.8 Concluding Remarks

Early mathematical models of cancer treated the tumor as an isolated population and focused on growth laws. These models clarified how tumor burden changes in time, but they did not account for the continuous interaction between the tumor and the immune system. Here we introduced two-dimensional tumor-immune models, which we called *Base Tumor-Immune Models*, to explicitly model tumor VS immune interaction. We then extended these tumor-immune models to include three standard cancer treatments. Surgery was modeled as an instantaneous removal of tumor mass, radiotherapy was modeled as tumor cell killing over a treatment window, and chemotherapy was modeled as a treatment that affects both tumor and immune cells. Writing each treatment scenario required expressing the resulting models as initial value problems (IVPs), which allowed us to compute and visualize solution curves in MATLAB.

In all of these two-dimensional tumor-immune-treatment models, exactly one long-term outcome was stable for any given parameter set: either the tumor persisted or the tumor was eliminated. A widely used research-level tumor-immune model introduced by Kuznetsov et al. [1] incorporates nonlinear immune stimulation and can exhibit *bistability*, in which both tumor-persistent and tumor-free equilibria are stable and separated by an unstable threshold. Using the Kuznetsov model as a base tumor-immune model is therefore a natural next step for students interested in richer tumor-immune dynamics.

We selected melanoma as the case study because melanoma is strongly immunogenic: immune activation, antigen presentation, and immune evasion are central to its progression. The model [8] represents these interactions using two compartments: the injection point compartment and the skin compartment. Using QSS approximations, we reduced the original seven-equation system to a four-equation model that retained the essential feedback among tumor cells, antigen presentation, antibodies, and effector cells. The reduced model was tractable enough to analyze equilibria, feasibility conditions, global asymptotic analysis. This kind of model reduction is useful for undergraduate learning and research. Starting from a published, complex system and simplifying it step-by-step

Table 4.3 Mapping the extended reduced tumor–immune model (4.27) to the updated 2011 Hallmarks of Cancer [19]. Each term corresponds to a biological process that reflects one of the ten biological capabilities. The use of Hill-type functions captures nonlinear saturation

Mapping the extended reduced model (4.27) with the 2011 hallmarks of cancer

$$\frac{dE}{dt} = p + \alpha_7 \frac{A}{A+k_1} E + \beta_6 A - \alpha_8 E, \quad \frac{dC}{dt} = \alpha_1 C\left(1 - \frac{C}{K}\right) - \alpha_3 \frac{A+k_2}{A+k_3} EC,$$

$$\frac{dA_b}{dt} = q + \alpha_{11} E - \alpha_{10} A_b, \quad \frac{dA}{dt} = \alpha_4 \alpha_3 \frac{A+k_2}{A+k_3} EC - (\alpha_5 + \beta_6 + \alpha_{12} A_b) A$$

Term	Hallmark / interpretation
$\alpha_1 C(1 - C/K)$	**Sustaining proliferative signaling** **Evading growth suppressors**
$-\alpha_3 \frac{A+k_2}{A+k_3} EC$	**Resisting cell death** Immune-mediated killing via a saturating Hill-type response; strong saturation reflects resistance to death
p	**Enabling replicative immortality** Constant immune source
$\alpha_7 \frac{A}{A+k_1} E$	**Avoiding immune destruction** Saturating Hill-type response limits activation at high antigen levels, which enables immune evasion
$\beta_6 A$	**Tumor-promoting inflammation** Antigen-driven activation can increase immune feedback
$-\alpha_8 E$	**Avoiding immune destruction** Effector decay decreases immune response
q	**Enabling replicative immortality** Constant antibody source
$+\alpha_{11} E - \alpha_{10} A_b$	**Tumor-promoting inflammation** Balance between antibody stimulation and decay influences continued inflammatory signaling
$+\alpha_4 \alpha_3 \frac{A+k_2}{A+k_3} EC$	**Genome instability and mutation** Generation of new antigens through ongoing tumor-immune interaction The saturating Hill-type response shows that repair capacity slows down at high antigen levels, allowing mutations to build up over time
$-(\alpha_5 + \beta_6 + \alpha_{12} A_b) A$	**Deregulating cellular energetics** Antigen clearance and use illustrate how cancer cells reprogram energy and resource use to support rapid growth

using QSS ideas provides a natural bridge between research papers and undergraduate studies and explorations.

Throughout the chapter, students practiced:

- translating biological mechanisms into ODE terms;
- incorporating treatment interventions (surgery, radiotherapy, and chemotherapy) into tumor-immune models;
- applying QSS reduction to eliminate fast variables to reduce system;
- verifying forward invariance of biologically meaningful regions;
- performing equilibrium and stability analysis, including Lyapunov arguments;
- using MATLAB simulations to visualize and validate model behavior (with and without treatments).

The *Hallmarks of Cancer* framework was used to analyze model structure. We mapped the terms in the 2D tumor-immune models to the six classical hallmarks and the reduced model to the 10 hallmarks. This mapping illustrates how each mathematical term represents a biological capability and how treatments influence the system dynamics. Students are encouraged to use the hallmarks (most recently updated in 2022 [20]) when analyzing tumor-immune models.

Appendix: Variables and Parameter Values Used in Simulations

Here, we summarize the variables and parameter values used in the MATLAB simulations throughout the chapter (Tables 4.4, 4.5, 4.6, and 4.7). All accompanying MATLAB codes are available in the project's GitHub repository [5].

Exercises

4.1 (Deriving von Bertalanffy Model from First Principles) In this activity, you will derive the Bertalanffy's model for the volume of a tumor, denoted by $V(t)$, where t is time (for example, in days).

The tumor grows because it gains material (cell division, proliferation) and loses material (cell death, metabolic costs). Hence, the rate of change of the tumor's volume can be expressed with the balance equation:

$$\frac{dV}{dt} = \text{(growth rate)} - \text{(loss rate)}.$$

Let's make the following assumptions based on biological principles:

(a) Suppose that the **growth** of the tumor (net gain of volume due to cell division) is limited by nutrient uptake through the tumor surface. This suggests that the **growth**

Table 4.4 Summary of variables and parameters for the phase-plane tumor-immune models used in Figs. 4.2 and 4.7. See Sect. 4.4

Symbol	Description	Value/model
State variables (both models)		
$C(t)$	Tumor population	—
$E(t)$	Effector immune population	—
Lotka-Volterra phase plane (Fig. 4.2)		
r	Tumor exponential growth rate	0.5
k	Tumor killing coefficient	0.3
s	Effector stimulation rate	0.4
d	Effector decay rate	0.2
Domain	Phase-plane range	$C \in [0, 7],\ E \in [0, 7]$
Kuznetsov Tumor-Immune model phase plane (Fig. 4.7)		
α	Tumor growth rate	1.636
β	Inverse carrying capacity	0.002
ρ	Immune stimulation rate	1.131
η	Half-saturation constant	20.19
μ	Tumor-immune interaction coefficient	0.00311
δ	Effector decay rate	0.3743
σ	Baseline immune influx	0.1181
Domain	Phase-plane range	$C \in [0, 500],\ E \in [0, 5]$

rate is proportional to the surface area S. Let $a > 0$ be a proportionality constant. Using the formulas for surface area $S = 4\pi r^2$ and volume $V = 4/3\pi r^3$, show that the growth rate term can be written as:

$$\text{growth rate} = a\, V^{2/3}.$$

(b) Suppose that the **loss** of tumor volume (cell death, metabolic costs) is **proportional to the tumor size (volume)** itself, hence

$$\text{loss rate} = b\, V,$$

for some $b > 0$ proportionality constant.

(c) Show that the tumor volume rate of change is given my the ODE:

$$\frac{dV}{dt} = aV^{2/3} - bV.$$

(d) This equation is a Bernoulli type of linear differential equation. It could be solved using the substitution $u = V^{1/3}$ or $V = u^3$. Use this substitution to show that the

Table 4.5 Summary of variables, parameters, and treatment settings for all evolution-plot (time-series) tumor-immune models used in Figs. 4.3, 4.4, 4.5, and 4.6. See Sect. 4.4.3

Symbol	Description	Value/model
State variables (all models)		
$C(t)$	Tumor population	—
$E(t)$	Effector immune population	—
Shared Tumor–Immune parameters		
r	Tumor logistic growth rate	0.8
K	Tumor carrying capacity	1.0
p	Immune killing efficiency	0.6
c	Effector stimulation rate	0.2
d	Natural effector decay rate	0.3
s_{pre}	Baseline immune influx	0.1
Cancer vs immunity without treatment (Fig. 4.3)		
p_{rec}	Weak immune killing	0.6
s_{rec}	Weak immune influx	0.1
p_{elim}	Strong immune killing	0.2
s_{elim}	Strong immune influx	0.5
Cancer vs immunity and surgery (Fig. 4.4)		
t_s	Surgery time	20
σ_{rec}	Tumor removed (recurrence)	0.6
σ_{elim}	Tumor removed (elimination)	0.6
$s_{\text{post, rec}}$	Post-surgery immune influx	0.2
$s_{\text{post, elim}}$	Post-surgery immune influx	0.5
Cancer vs immunity and radiotherapy (Fig. 4.5)		
t_r	Radiation start time	20
T_r	Radiation duration	10
ρ_{rec}	Radiation strength (recurrence)	0.6
ρ_{elim}	Radiation strength (elimination)	0.8
$s_{\text{post, rec}}$	Immune recovery post-radiation	0.2
$s_{\text{post, elim}}$	Immune recovery post-radiation	0.5
Cancer vs immunity and chemotherapy (Fig. 4.6)		
t_c	Chemotherapy start time	20
T_c	Chemotherapy duration	10
γ_C	Tumor death induced by chemo	0.6
γ_E	Effector death induced by chemo	0.1
s_{post}	Immune activation after chemo	0.4
Initial conditions (applied to all)		
$C(0)$	Initial tumor level	0.2
$E(0)$	Initial effector level	0.1
t	Simulation interval	$[0, 60]$

Table 4.6 Parameter magnitudes and associated timescales in Pennisi's full melanoma–immune model. All rates are given in units of Δt^{-1} with $\Delta t = 8$ hours

Symbol	Biological process	Value (Δt^{-1})	Timescale $\tau = 1/r$ (Δt)	Timescale (hours)	Classification
$\alpha_1, \alpha_2, \alpha_4, \alpha_5$	Tumor and antigen dynamics	0.01–0.22	4.5–100	$\approx$36–800	Slow (tumor/antigen)
α_6	Naive T-cell activation in skin (N)	0.10	10	80	Fast
α_7	Activation of effectors in skin (E_s)	0.095	$\approx$10.5	$\approx$84	Fast
α_8	Additional local immune process in skin	0.03–0.10	10–33.3	$\approx$80–267	Fast / intermediate
α_{10}	Antibody dynamics in skin (A_s)	0.033	$\approx$30.3	$\approx$242	Fast/intermediate
h	Antigen supply/naive T-cell source term	0.01	100	800	Slow
α_{11}	Migration from injection site to skin	0.009	$\approx$111.1	$\approx$889	Slowest

Table 4.7 Variables, parameters, and initial conditions for the reduced tumor-immune model, Sect. 4.5

Symbol	Description	Value
State variables		
$E(t)$	Effector immune cells	–
$A_b(t)$	Antibody (or effector-associated) population	–
$C(t)$	Tumor cell population	–
$A(t)$	Antigen concentration	–
Model parameters		
p	Constant input of effector cells (immunotherapy)	0.1
q	Constant input of antibody	0.05
α_1	Tumor logistic growth rate	1.0
K	Tumor carrying capacity	1.0
β_3	Tumor-killing efficiency of effector cells	2.5 (tumor-free), 0.5 (tumor-persistent)
α_4	Antigen release scaling	0.6
α_5	Natural antigen clearance rate	0.4
β_6	Antibody-mediated antigen clearance rate	0.2
β_8	Effector decay rate	0.3
β_{10}	Antibody decay rate	0.25
k_2	Saturation parameter in $G(A)$	0.4
k_3	Saturation parameter in $G(A)$	0.3
Initial conditions and simulation window		
$E(0)$	Initial effector level	0.5
$A_b(0)$	Initial antibody level	0.2
$C(0)$	Initial tumor level	0.6
$A(0)$	Initial antigen level	0.2
t	Simulation interval	[0, 100] (tumor-free), [0, 200] (tumor-persistent)

solution is

$$V(t) = \left[\frac{a}{b} + \left(V_0^{1/3} - \frac{a}{b}\right) e^{-bt/3}\right]^3,$$

where V_0 is the volume of the tumor at initial time t_0.

4.2 (Deriving Gompertz Model from First Principles) In this activity, you will derive the Gompertz's model for the volume of a tumor, denoted by $V(t)$, where t is time (for example, in days).

A tumor receives nutrients and oxygen by diffusion from outside its boundary. Assuming the tumor is approximately spherical, the cells near the surface receive more nutrients than the cells deep inside. As the tumor grows, nutrients penetrate a smaller proportion of the interior, so the *proliferative region* becomes thinner over time.

Let $r(t)$ denote an *effective proliferative depth*: the distance from the surface into which nutrients can successfully diffuse and support active growth. As the tumor expands and becomes denser, $r(t)$ decreases with time, reducing the fraction of proliferating cells.

(a) Explain why the two differential equations

$$\frac{dr}{dt} = -\beta r(t), \qquad \frac{dV}{dt} = r(t)\,V(t) \tag{4.28}$$

describe the above biological assumptions.

(b) Reduce this system to one equation. Use the following relationship:

$$\frac{1}{V}\frac{dV}{dt} = r = -\frac{1}{\beta}\frac{dr}{dt}$$

and integrate both sides with respect to t. Obtain $r = \alpha - \beta \ln V$ and substitute back in Eq. (4.28) for the rate of change of V to get the **Gompertz model**

$$\frac{dV}{dt} = (\alpha - \beta \ln V)\,V. \tag{4.29}$$

(c) Show that Eq. (4.29) can be written in the following format

$$\frac{dV}{dt} = aV \ln\left(\frac{1}{bV}\right), \tag{4.30}$$

where $a = \beta$ and $1/b = \exp(\alpha/\beta)$.

(d) Solve equation (4.30) to obtain the closed form solution

$$V(t) = \frac{1}{b} \exp\left(\ln(bV_0)e^{-at}\right),$$

where V_0 is the volume at the initial time t_0.

Remark To solve equation (4.30), use the substitution $u = \ln(bV)$ and the fact that $du = \frac{dV}{V}$ to obtain after separation of variables that

$$\int \frac{du}{u} = -\int a dt.$$

After integrating, substitute $u = \ln(bV)$ back and express V.

4.3 (Cancer Growth Models) Classical cancer growth models are in Table 4.1.

(a) Solve each ODE using separation of variables. Note that for the generalized von Bertalanffy model, after separating the variables, you should use the substitution $u = (bC)^r$. As a result your solution will be similar to the solution of the logistic equation and maybe solved by the method of partial fractions.
(b) Verify that your solutions match the corresponding closed-form expressions in the table. Some algebraic manipulations may be required.
(c) For each model, compute $\lim_{t\to\infty} C(t)$ and interpret the result in terms of tumor dynamics, that is, unlimited growth, bounded growth, saturation.

4.4 (Cancer Growth Models) Consider the cancer growth models with parameters $a = 0.3$, $b = 0.01$, $r = 2$, and initial tumor size $C_0 = 1$ (in arbitrary volume units).

(a) Sketch or plot the cancer growth curves $C(t)$ for $0 \leq t \leq 50$.
(b) For the three bounded models (Logistic, Gompertz, von Bertalanffy), compute the time to reach 90% of the asymptotic size, that is,

$$t_{0.9} = \min\{\, t \geq 0 : C(t) = 0.9K \,\},$$

where $K = 1/b$ is the carrying capacity. For the exponential model (which has no finite asymptote), compute the time to reach the same reference value $C(t) = 0.9K$.
(c) Identify and briefly explain which models:

- predict unlimited growth,
- predict bounded logistic-type saturation,
- predict sub-exponential slowing,
- and which include an adjustable shape parameter that changes the curvature of the trajectory without changing the value of the asymptote.

(d) By looking at the plots of the solutions, discuss how the choice of model affects cancer growth predictions in cancer treatments.

4.5 (Cancer Growth Models) For each of the cancer growth models in Table 4.1:

(a) Determine the *carrying capacity* K, i.e., the asymptotic tumor size $K = \lim_{t\to\infty} C(t)$.
(b) Show how K can be obtained directly by solving for steady states of the ODE $\frac{dC}{dt} = 0$, and discuss whether these equilibria are stable.
(c) Which models have a finite positive carrying capacity, no finite carrying capacity, or a carrying capacity that depends on parameters in a nontrivial way?

(d) Interpret your results in a biological context. What kind of cancer growth dynamics is each model capturing in the context of carrying capacity?

4.6 (Chemotherapy as Piecewise Autonomous Systems) In Sect. 4.4 Example 3, chemotherapy is applied over the fixed interval $[t_c, t_c + T_c]$ and affects both tumor and effector cells. The model is

$$\begin{cases} \frac{dC}{dt} = r\,C\left(1 - \frac{C}{K}\right) - p\,E\,C - \gamma_C\,C\,u(t), \\ \frac{dE}{dt} = s + c\,C - d\,E - \gamma_E\,E\,u(t), \end{cases}$$

where $u(t)$ is the indicator function

$$u(t) = \begin{cases} 0, & 0 \le t < t_c, \\ 1, & t_c \le t \le t_c + T_c, \\ 0, & t > t_c + T_c. \end{cases}$$

The terms $-\gamma_C Cu(t)$ and $-\gamma_E Eu(t)$ represent chemotherapy-induced-cancer-cell and effector-cell death, respectively.

(a) Rewrite the non-autonomous system above as *three autonomous IVPs* corresponding to (i) before chemotherapy, (ii) during chemotherapy, and (iii) after chemotherapy. Clearly identify each system and its initial condition. (See Sect. 4.4 Example 2 for a demonstration.)
(b) Explain how this piecewise-autonomous formulation simplifies the analysis of the system compared to keeping the non-autonomous $u(t)$ term.
(c) Discuss how the additional death terms γ_C and γ_E may affect the qualitative dynamics of the tumor and immune populations in the middle time interval compared to the other two intervals.

4.7 (Equilibria and Surgery) Consider the tumor–immune system with a surgical intervention at time $t = t_s$,

$$\frac{dC}{dt} = rC\left(1 - \frac{C}{K}\right) - pEC - \sigma C\,\delta(t - t_s), \qquad \frac{dE}{dt} = s + cC - dE.$$

(a) Ignore the delta term and compute the equilibrium points (C^*, E^*) of the underlying autonomous system.

(b) Explain why the surgery term does not affect the equilibrium points, even though it affects the solution trajectory.
(c) Use MATLAB to simulate the system and produce a phase plane to describe qualitatively what surgery does to the phase plane trajectory of the system but not to the equilibria.

4.8 (Equilibria and Radiation Therapy) Consider the tumor–immune system with radiation therapy modeled by

$$\frac{dC}{dt} = rC\left(1 - \frac{C}{K}\right) - pEC - \rho C\,u(t), \qquad \frac{dE}{dt} = s + cC - dE,$$

where $u(t) = 1$ during the treatment window and $u(t) = 0$ otherwise.

(a) Compute the equilibrium $(C^*_{\text{off}}, E^*_{\text{off}})$ before treatment ($\rho = 0$).
(b) Compute the equilibrium $(C^*_{\text{on}}, E^*_{\text{on}})$ during treatment ($\rho > 0$).
(c) Compare C^*_{on} and C^*_{off} and explain why they differ. What happens to the equilibria after treatment ends?

4.9 (Equilibria and Chemotherapy) Consider the tumor–immune system with chemotherapy modeled by

$$\frac{dC}{dt} = rC\left(1 - \frac{C}{K}\right) - pEC - \gamma_C C\,u(t), \qquad \frac{dE}{dt} = s + cC - dE - \gamma_E E\,u(t),$$

where $u(t) = 1$ during treatment and $u(t) = 0$ otherwise.

(a) Compute the equilibrium $(C^*_{\text{off}}, E^*_{\text{off}})$ before treatment ($\gamma_C = \gamma_E = 0$).
(b) Compute the equilibrium $(C^*_{\text{on}}, E^*_{\text{on}})$ during treatment ($\gamma_C > 0$, $\gamma_E > 0$).
(c) Compare the tumor equilibria before and during treatment and explain how immune suppression (γ_E) can influence whether the tumor rebounds after treatment ends.

4.10 (Combination Therapy) Consider the tumor–immune model with surgery, radiation, and chemotherapy introduced in Sect. 4.4 Example 4:

$$\begin{cases} \dfrac{dC}{dt} = r\,C\left(1 - \dfrac{C}{K}\right) - p\,E\,C - \rho\,C\,u_r(t) - \gamma_C\,C\,u_c(t), \\ \dfrac{dE}{dt} = s + c\,C - d\,E - \gamma_E\,E\,u_c(t), \end{cases}$$

where surgery at $t = t_s$ removes a fraction σ of the tumor, radiation is active on $[t_r,\ t_r+T_r]$, and chemotherapy is active on $[t_c,\ t_c + T_c]$. The indicator functions $u_r(t)$ and $u_c(t)$ are defined by

$$u_r(t) = \begin{cases} 1, & t_r \le t \le t_r + T_r, \\ 0, & \text{otherwise}, \end{cases} \qquad u_c(t) = \begin{cases} 1, & t_c \le t \le t_c + T_c, \\ 0, & \text{otherwise}. \end{cases}$$

and the surgery impulse is $C(t_s^+) = (1 - \sigma)C(t_s^-)$ and $E(t_s^+) = E(t_s^-)$.

(a) Identify all time intervals in which the right-hand side of the system is autonomous. Clearly label each interval (e.g., before surgery, during radiation, overlap of radiation and chemotherapy, after treatment).
(b) For each interval identified in part (a), rewrite the system as an autonomous IVP with constant parameters and appropriate initial conditions determined by previous intervals.
(c) Suppose the treatments do not overlap ($t_s < t_r < t_r + T_r < t_c < t_c + T_c$). How many autonomous IVPs are needed to fully represent the treatment timeline?
(d) Simulate the combo therapy using MATLAB

4.11 (Model Reduction) Consider the following within-host virus-immune interaction model with five state variables: $T(t), I(t), V(t), E(t), A(t)$, representing uninfected target cells, infected cells, virus particles, cytotoxic effector cells, and activated immune cells, respectively. The governing equations are

$$\begin{cases} \dfrac{dT}{dt} = \lambda - d_T T - \beta T V, \\ \dfrac{dI}{dt} = \beta T V - \delta I - k E I, \\ \dfrac{dV}{dt} = pI - cV, \\ \dfrac{dE}{dt} = s + \alpha A - d_E E, \\ \dfrac{dA}{dt} = rV - \gamma A. \end{cases}$$

Here λ is the production rate of target cells and d_T their death rate, β is the infection rate, δ the infected cell death rate, p the viral production rate, and c the viral clearance rate. The immune response has baseline production s, stimulation rate α, activation rate r, and clearance rate γ.

(a) **Time-scale identification.** Biological data suggest that the activated immune cell population $A(t)$ responds and clears on a much faster time scale (hours) than the other variables (days). Explain how this motivates a quasi-steady-state approximation for $A(t)$. Write down the quasi-steady-state assumption for $A(t)$.

(b) **Solving for the fast variable.** Set $\frac{dA}{dt} = 0$ and solve for $A^*(V)$ as a function of V. Substitute $A^*(V)$ into the equation for $E(t)$. *Hint:* $0 = rV - \gamma A^* \Rightarrow A^*(V) = \frac{r}{\gamma}V$.

(c) **Deriving the reduced system.** Write the reduced four-variable system in $T(t)$, $I(t)$, $V(t)$, and $E(t)$. Identify which terms in the E-equation are modified as a result of this reduction. Expected form is:

$$\begin{cases} \dfrac{dT}{dt} = \lambda - d_T T - \beta T V, \\ \dfrac{dI}{dt} = \beta T V - \delta I - kEI, \\ \dfrac{dV}{dt} = pI - cV, \\ \dfrac{dE}{dt} = s + \alpha \dfrac{r}{\gamma} V - d_E E. \end{cases}$$

(d) **ODE toolbox.** Perform a local stability analysis of the infection-free equilibrium.

(e) **Model reflection.** Explain the biological meaning of replacing $A(t)$ by its steady state $A^*(V)$. Discuss how this reduction simplifies mathematical analysis (for example, equilibrium computation, stability, Lyapunov analysis).

4.12 (Model Reduction) Consider the following predator-prey model with three state variables: $R(t)$, $N(t)$, $P(t)$, representing the resource concentration (e.g., nutrient), prey population (e.g., plankton), and predator population (e.g., fish), respectively. The governing equations are

$$\begin{cases} \dfrac{dR}{dt} = S - \mu R - aRN, \\ \dfrac{dN}{dt} = baRN - dN - cNP, \\ \dfrac{dP}{dt} = ecNP - mP. \end{cases}$$

Here S is the nutrient input rate, μ is the natural decay rate of the resource, and a is the prey uptake rate. The parameter b represents the conversion efficiency of resource to prey biomass, while d is the prey death rate. Predation occurs at rate c, and e is the conversion efficiency from prey to predator biomass. Finally, m denotes the predator death rate.

(a) **Time-scale identification.** In many ecological systems, resources (e.g., nutrients) equilibrate rapidly compared to biological populations. Explain why $R(t)$ is a natural candidate for a *fast variable*. Write down the quasi-steady-state (QSS) assumption for $R(t)$.

(b) **Solving for the fast variable.** Set $\frac{dR}{dt} = 0$ and solve for $R^*(N)$ in terms of the prey density N. Start from the first equation $S - \mu R - aRN = 0$ and show that $R^*(N) = \frac{S}{\mu + aN}$.

(c) **Deriving the reduced system.** Substitute $R^*(N)$ into the prey equation to eliminate R. Write the reduced two-variable system in $N(t)$ and $P(t)$. *Expected form:*

$$\begin{cases} \frac{dN}{dt} = ba\frac{S}{\mu+aN}N - dN - cNP, \\ \frac{dP}{dt} = ecNP - mP. \end{cases}$$

(d1) **ODE toolbox.** Sketch the nullclines of the reduced system and determine the condition for coexistence equilibrium of prey and predator.

(d2) **ODE toolbox.** Use `pplane.jar` to visualize trajectories, direction fields, sketch nullclines, and analyze stability of the equilibria of the two-dimensional ODE system.

(e) **Model reflection.** Explain the biological meaning of replacing $R(t)$ by its steady state $R^*(N)$. Discuss why the reduced model resembles a predator-prey system with a nonlinear (saturating) prey growth term.

4.13 (Exploring an Alternative Reduction) In the QSS reduction Sect. 4.5.2, the antigen-dependent activation term

$$H(A) = \alpha_7 \frac{A}{A + k_1} E$$

represents a simplified version of the fast expression $\alpha_7(A_s/(A_s + k_1))E_s$ obtained from the full Pennisi model (4.8)–(4.14). In our derivation we replaced the detailed dependence on the fast variables A_s and E_s by a Hill-type function of the antigen A, motivated by the biological role of antigen as the primary driver of effector stimulation. As an alternative modeling choice, suppose we instead approximate $A_s \approx k_A A_b$ and replace the stimulation term by a Hill function of the antibody population,

$$\alpha_7 \frac{A_b}{A_b + k_1} E.$$

Similarly, consider replacing the antigen-modulated killing term in (4.21)–(4.22) by a function of A_b,

$$\alpha_3 \frac{A_b + k_2}{A_b + k_3} EC.$$

1. Using the QSS relations $A_s \approx k_A A_b$ and $E_s \approx k_E(\alpha_{11} E + \beta_6 A)$, explain why $H(A_b)$ and $G(A_b)$ are mathematically plausible alternative approximations.
2. Discuss whether these antibody-based functions agree with the biological roles of antigen and antibody in the Pennisi model. In particular, which population (antigen A or antibody A_b) provides the primary signal for effector activation?
3. Compare the resulting reduced models obtained using $H(A)$ versus $H(A_b)$. How do the feedback pathways differ? Which choice yields a model that more accurately reflects the qualitative behavior of the original system?

4.14 (Hallmarks-to-Model Mapping) In Sect. 4.7, we mapped the six hallmarks of cancer to Base Tumor-Immune Model 2 and the reduced tumor-immune model to the updated list of hallmarks. Consider the full Pennisi model (4.8)–(4.14), and the reduced model (4.19)–(4.22). In this exercise you will relate the biological processes represented in these equations to the Hallmarks of Cancer (2011 update).

1. **Full Pennisi model.** For each term in Eqs. (4.8)–(4.14), identify the biological process represented and map it to one or more of the 2011 hallmarks (six original hallmarks, two enabling characteristics, and two emerging hallmarks). Provide a short justification for each mapping. Use the structure of the table in this chapter as a guide.
2. **Reduced model.** Repeat the mapping for the reduced system (4.19)–(4.22). Compare your results with Part (1):

   - Which hallmarks are still represented explicitly?
   - Which hallmarks become implicit or disappear in the reduced formulation?
   - How does reducing biological detail change the mathematical expression of the hallmarks?

3. **Reflection.** Write a short paragraph explaining how model reduction via QSS affects the ability of an ODE system to encode biological hallmarks. Comment on which biological features are preserved, which are lost, and which become more transparent mathematically.

List of Notations for Chap. 4

Symbol	Description/meaning	Units
State Variables		
$C(t)$	Tumor cell population (density or nondimensionalized)	Cells or nondim.
$E(t)$	Effector (cytotoxic) immune cells	Cells or nondim.
$A_b(t)$	Circulating antibodies	Nondimensional
$A(t)$	Tumor-released antigen	Nondimensional
Core Model Parameters		
r, α_1	Tumor growth rate (logistic)	day^{-1}
K	Tumor carrying capacity	Nondimensional
p, α_3, β_3	Immune-mediated tumor killing strength	Day^{-1}
c, α_7	Antigen-driven stimulation of effector cells	Day^{-1}
d, α_8	Natural effector cell decay rate	Day^{-1}
q	Baseline antibody production rate	Day^{-1}
α_{10}	Antibody decay rate	Day^{-1}
α_{11}	Antibody production stimulated by effector cells	Day^{-1}
α_{12}	Antibody–antigen binding / neutralization rate	Day^{-1}
β_6	Antigen feedback on effector activation	day^{-1}
k_1, k_2, k_3	Antigen-processing saturation parameters	Nondimensional
Treatment Parameters		
σ	Fraction of tumor removed at surgery time t_s	—
t_s	Time of surgical intervention	Days
ρ	Radiation-induced tumor death rate during treatment	Day^{-1}
t_r	Radiation start time	Days
T_r	Duration of radiation exposure	Days
γ_C	Chemotherapy-induced tumor cell death rate	Day^{-1}
γ_E	Chemotherapy-induced effector cell death rate	Day^{-1}
t_c	Chemotherapy start time	Days
T_c	Duration of chemotherapy exposure	Days

References

1. Kuznetsov, V. A., Makalkin, I. A. , Taylor, M. A., and Perelson, A. S., Nonlinear dynamics of immunogenic tumors: parameter estimation and global bifurcation analysis, *Bulletin of Mathematical Biology*, vol. 56, no. 2, pp. 295–321, (1994).

2. Rockne, R.C., Hawkins-Daarud, A., Swanson, K.R., *et al.* The 2019 mathematical oncology roadmap. *Physical Biology*. 16(4):041005. (2019) https://doi.org/10.1088/1478-3975/ab1a09.
3. Belkhir, S., Thomas, F., Roche, B., Darwinian approaches for cancer treatment: Benefits of mathematical modeling. *Cancers*. 13(17):4448. (2021)) https://doi.org/10.3390/cancers13174448.
4. de Pillis, L. G., Radunskaya, A. E. , Modeling Tumor–Immune Dynamics, in *Mathematical Models of Tumor–Immune System Dynamics*, A. Eladdadi, P. Kim, and D. Mallet, Eds. Springer, New York, NY, pp. 59–108. (2014) https://doi.org/10.1007/978-1-4939-1793-8_4.
5. Savatorova, V., Panayotova, I., Hallare, M. (2025). Mathematical-modeling-in-Life-Sciences-A-Practical-Project-Based-Approach [Source code]. GitHub. https://github.com/ViktoriaLS/Mathematical-modeling-in-Life-Sciences-A-Practical-Project-Based-Approach
6. Hanahan, D., Weinberg, R.A., The hallmarks of cancer. *Cell*. 2000;100(1):57–70.
7. Rui, R., Zhou, L., and He, S., Cancer immunotherapies: advances and bottlenecks, *Frontiers in Immunology*, vol. 14, article 1212476, (2023) https://doi.org/10.3389/fimmu.2023.1212476.
8. Pennisi, M. , A Mathematical Model of Immune-System-Melanoma Competition, *Computational and Mathematical Methods in Medicine*, vol. 2012, Article ID 850754, 13 pp., (2012). https://doi.org/10.1155/2012/850754.
9. Zill, D. G., *Advanced Engineering Mathematics with WebAssign*. 7th ed., Jones & Bartlett Learning, (2020).
10. Strogatz, S. H., *Nonlinear Dynamics and Chaos: With Applications to Physics, Biology, Chemistry, and Engineering*. 2nd ed., Westview Press, (2014).
11. Stepien, T. L., Kostelich, E. J., and Kuang, Y., Mathematics + Cancer: An Undergraduate "Bridge" Course in Applied Mathematics, *SIAM Review*, 62(1):244–263, (2020).
12. Bryan, K. *Differential Equations: A Toolbox for Modeling the World*. SIMIODE / QUBES; (2024). Available at: https://qubeshub.org/community/groups/simiode/textbook. ISBN: 978-1-63877-937-7 (PDF), 979-8-89372-019-8 (HTML).
13. Bekker, R. A., Kim, S., Pilon-Thomas, S., and Enderling, H., Mathematical modeling of radiotherapy and its impact on tumor interactions with the immune system, *Neoplasia*, 28:100796, (2022).
14. Hallare, M. , and Santamore, J. M., Expanding ODE examples: Introducing gene regulation dynamics through Hill functions, *CODEE Journal*, vol. 19, no. 1, Article 6, (2025). Available: https://scholarship.claremont.edu/codee/vol19/iss1/6.
15. Shain, A.H., Bastian, B.C., From melanocytes to melanomas. *Nature Reviews Cancer*. 16(6):345–358. (2016) https://doi.org/10.1038/nrc.2016.37.
16. Wilkie, K.P., A review of mathematical models of cancer–immune interactions in the context of melanoma. *Mathematical Medicine and Biology*. 25(1):1–17. (2008) https://doi.org/10.1093/imammb/dqm006.
17. Khalil, H.K., *Nonlinear Systems*, Third Edition. Prentice Hall; (2002). ISBN: 0-13-067389-7.
18. Murray, J.D., *Mathematical Biology*, Third Edition, Springer, (2002).
19. Hanahan, D., Weinberg, R.A., Hallmarks of cancer: The next generation. *Cell*. 144(5):646–674. (2011)
20. Hanahan, D., Hallmarks of cancer: New dimensions. *Cancer Discovery*. 12(1):31–46.(2022) https://doi.org/10.1158/2159-8290.CD-21-1059.